T0205424

Smart Innovation, Systems and Technologies

Volume 87

Series editors

Robert James Howlett, Bournemouth University and KES International,
Shoreham-by-sea, UK
e-mail: rjhowlett@kesinternational.org

Lakhmi C. Jain, University of Canberra, Canberra, Australia;
Bournemouth University, UK;
KES International, UK
e-mails: jainlc2002@yahoo.co.uk; Lakhmi.Jain@canberra.edu.au

The Smart Innovation, Systems and Technologies book series encompasses the topics of knowledge, intelligence, innovation and sustainability. The aim of the series is to make available a platform for the publication of books on all aspects of single and multi-disciplinary research on these themes in order to make the latest results available in a readily-accessible form. Volumes on interdisciplinary research combining two or more of these areas is particularly sought.

The series covers systems and paradigms that employ knowledge and intelligence in a broad sense. Its scope is systems having embedded knowledge and intelligence, which may be applied to the solution of world problems in industry, the environment and the community. It also focusses on the knowledge-transfer methodologies and innovation strategies employed to make this happen effectively. The combination of intelligent systems tools and a broad range of applications introduces a need for a synergy of disciplines from science, technology, business and the humanities. The series will include conference proceedings, edited collections, monographs, handbooks, reference books, and other relevant types of book in areas of science and technology where smart systems and technologies can offer innovative solutions.

High quality content is an essential feature for all book proposals accepted for the series. It is expected that editors of all accepted volumes will ensure that contributions are subjected to an appropriate level of reviewing process and adhere to KES quality principles.

More information about this series at http://www.springer.com/series/8767

Fábio Romeu de Carvalho
Jair Minoro Abe

A Paraconsistent
Decision-Making Method

 Springer

Fábio Romeu de Carvalho
Paulista University, UNIP
São Paulo
Brazil

Jair Minoro Abe
Paulista University, UNIP
São Paulo
Brazil

Additional material to this book can be downloaded from http://extras.springer.com.

ISSN 2190-3018 ISSN 2190-3026 (electronic)
Smart Innovation, Systems and Technologies
ISBN 978-3-030-08919-1 ISBN 978-3-319-74110-9 (eBook)
https://doi.org/10.1007/978-3-319-74110-9

*Dedicated to Professor Newton
C. A. da Costa, teacher and friend.*

Foreword

Nonclassical logic is a logic with some features which are different from those in classical logic. Consequently, nonclassical logics have been applied to applications for some areas, in particular, computer science and engineering. In fact, there are many logical systems classified as nonclassical logics.

In general, real systems face contradiction for many reasons. Classical logic cannot properly handle contradiction, and it is not regarded as an ideal system. Logics which are capable of dealing with contradiction are called paraconsistent logics. Now, the importance of paraconsistent logics is certainly recognized both in logic and computer science.

In fact, many different paraconsistent logics have been developed. One of the important paraconsistent logics was proposed by Newton da Costa in the 1950s. He is a leading person in the area. One of the authors of the present book, Jair Minoro, is da Costa's student and completed Ph.D. thesis under him on annotated logics in 1992.

Annotated logic was developed by Subrahmanian to provide a theoretical foundation for paraconsistent logic programming in 1987. Later, the logic has been studied by many including da Costa and Abe. The distinguished features of annotated logic are as follows: (1) it has a firm logical foundation, and (2) it is suitable for practical applications. On these grounds, annotated logic can be seen as an interesting paraconsistent logic.

In the 1990s, I also studied annotated logics by myself in connection with AI applications. I met Abe in 1997. Since then, I worked with him and wrote many papers. Recently, I also published a book *Introduction to Annotated Logics* with Abe and Kazumi Nakamatsu by Springer. We are now working on several books on annotated logics for applications.

The present book is concerned with applications of annotated logics to engineering. In Chap. 1, the authors say: "the object of this work is to present the reader with the principles of the annotated paraconsistent logics and their application in decision-making, mainly, in Production Engineering: the Paraconsistent Decision Method-PDM, which is based on the para-analyzer algorithm."

They provide the theoretical foundation for the PDM by using the paraconsistent annotated evidential logic Eτ, which is a version of annotated logics. Eτ seems to be appropriate for engineering applications since it captures several types of information including contradiction and incompleteness in real problems.

I can find several merits of the book by reading it. First, it is easy even for beginners to understand it. Namely, it starts with introductory chapters on theoretical aspects and goes on to chapters on engineering applications. Second, it is also regarded as a reference for experts. They can learn some aspects of annotated logics. Third, it covers many applications in some areas using annotated logic. Also, their approaches are carefully compared with others in the literature to defend their advantages. Finally, it includes useful appendix and references. They appear to be helpful for the readers.

I believe that the present book is significant in that it reveals the approach of annotated logic to engineering applications. The book also suggests many possibilities of annotated logics beyond engineering, which should be worked out.

I conclude that the readers will be able to understand the broad applicability of annotated logic.

Kawasaki, Japan   Seiki Akama
June 2017

Preface

At the dusk of the past century and at the dawn of this one, Computing in general (including the Information Systems, Artificial Intelligence, Robotics, and Automation, among others) goes through a real revolution, never seen before. The paradigm not only of knowledge but also of technology and its applications suffered radical changes.

Operational Research has been a very broad and inexhaustible subject. Hundreds of researchers all over the world have dedicated to this issue, which progresses daily. To have an idea of its dimension, there is a world conference—*European Conference on Operational Research, EURO,* which is annually held and, in July 2010, held its 24th edition, in Lisbon. In 2009, it was held in Bonn, Germany, where we were together with other 2,221 researchers from 72 countries.

Within Operational Research, the study of decision-making is inserted. A lot has been researched about this subject, several decision-making methods have been developed, but until today, none has managed to end the subject and, we believe, none will be able to do so. A fast Internet browsing may show how much is researched and how much is published about the so-called *Decision Support Systems, DSS.* They constitute a class of information systems (including, but not limited to computational systems), which support decision-making activities in the organizations and the businesses.

Moreover, it is in this area that we ventured, seeking to provide one more contribution to the scientific world, developing a new decision method substantiated on a logic which is alternative to the classical one, recently discovered, the paraconsistent annotated evidential logic. We named it Paraconsistent Decision-Making Method, PDM.

It is worth to highlight that it was a Brazilian logician; Newton C. A. da Costa is among the pioneers who developed the first paraconsistent systems in all logical levels in 1958. Others pioneers were the Polish logician J. Łukasiewicz and S. Jaśkowski, and the Russian logician N. A. Vasiliev.

Decision-Making with Paraconsistent Annotated Logic Tools

da Costa developed a family of paraconsistent logics, the C_n propositional systems, the corresponding predicate calculi, and higher order logic (in the form of set theory), containing in this way all the common logical levels. Regarding this theme, da Costa has lectured in all the countries of South and North America and some countries in Europe.

He received, among several distinctions, the Moinho Santista Award on Natural Sciences (1994), the Jabuti Award on Natural Sciences (1995), the "Nicolaus Copernicus" Scientific Merit Medal of the University of Torun, Poland (1998). He is a full member of the International Institute of Philosophy of Paris, the first Brazilian person to belong to this institution. He is also Emeritus Professor of Campinas State University.

We believe there is not a reference in the literature that gives the reader a proper comprehension of the themes related to this logic, which we have discussed in the several scientific meetings we have participated. With this work, we intend to provide a contribution in this sense, disseminating this new logic class, the paraconsistent logics, and showing how they may be utilized in decision-making, especially when the database we have is provided with inconsistencies and imprecisions.

Therefore, the object of this work is to present the reader with the principles of the annotated paraconsistent logics and their application in decision-making, mainly, in Production Engineering: the Paraconsistent Decision-Making Method— PDM, which is based on the para-analyzer algorithm. Besides that, a comparison of the PDM with the statistical method is made, as well as with a simplified version of the fuzzy decision method. Examples of practical applications are thoroughly developed and discussed, with numerical applications, tables, and charts.

The theoretical foundation for the PDM is the paraconsistent annotated evidential logic Eτ maximization and minimization rules. These rules are applied to the degrees of favorable evidence or degrees of belief (a) and the degrees of contrary evidence or degrees of disbelief (b), the compose the so-called annotation constants: $\mu = (a; b)$. This application is performed using operators and may be done so in two different ways.

(1) Conducting the *maximization of the degrees of evidence* of a set of annotations, in order to seek the best favorable evidence (**highest** value of the degree of favorable evidence a) and the worst contrary evidence (**highest** value of the degree of contrary evidence b). This maximization is made by an Eτ logic operator, designated by **OR** (conjunction). For the case of a set of only two annotations, the application of this operator is as follows:

$$\mathbf{OR}\ \{(a_1; b_1), (a_2; b_2)\} = (\max\ \{a_1, a_2\};\ \max\ \{b_1, b_2\})$$

For the minimization, we do the opposite: we seek the worst favorable evidence (**lowest** value of the degree of favorable evidence a) and the best contrary evidence (**lowest** value of the degree of contrary evidence b). The operator that executes it is designated by **AND** (disjunction).

$$\textbf{AND } \{(a_1; b_1), (a_2; b_2)\} = (\min\{ a_1, a_2\}; \min\{b_1, b_2\})$$

(2) Performing the *maximization (or the minimization) of the degree of certainty* (H = $a - b$) of the set of annotations, a degree that, in a certain way, translates how much the information contained in this set enable to infer for the veracity or the falsity of the premise.

The maximization of the degree of certainty (H) is obtained seeking the best favorable evidence (**highest** value of the degree of favorable evidence *a*) and the best contrary evidence (**lowest** value of the degree of contrary evidence *b*). This maximization is made by an Eτ LOGIC OPERATOR, designated by **MAX and that, in this book, will be called maximizing.**

$$\textbf{MAX } \{(a_1; b_1), (a_2; b_2)\} = (\max\{a_1, a_2\}; \min\{b_1, b_2\})$$

Analogously, minimization seeks the worst favorable evidence (**lowest** value of the degree of favorable evidence *a*) and the worst contrary evidence (**highest** value of the degree of contrary evidence *b*). This minimization is made by the **MIN** operator that will be called **minimizing.**

$$\textbf{MIN } \{(a_1; b_1), (a_2; b_2)\} = (\min\{a_1, a_2\}; \max\{b_1, b_2\})$$

Therefore, we observe that there are two ways to apply the maximization and minimization rules of the logic Eτ. In some aspects, one has advantages over the other; in others, disadvantages. For example, the first way enables a better identification of the existent inconsistencies in the database, but on the other hand, the second one is more intuitive and leads to more predictable and coherent results.

In this work, we will opt for the second manner, that is, for the **MAX** and **MIN** operators. The decisions will be made based on the application of the so-called min-max rule, or optimistic decision rule, once it minimizes the best results.

For the execution of the operations demanded by the method, in Chap. 5, we developed a calculation program based on the Excel spreadsheet, which was named Calculation Program for the Paraconsistent Decision Method, CP of the PDM.

In Chap. 9, a discussion is established about two ways to interpret the maximization and minimization, enabling a comparison between them.

There are five appendices that accompany this book, with data and solutions for the several items that are presented and analyzed.

For each appendix, there are two versions: a blocked one (but not hidden), which leaves only the cells related to the data input of each analysis free for the reader to alter, although it shows the other ones, including the formulas; and a free one, which gives the reader the possibility to alter whatever they consider necessary.

This concern resulted from the possibility of a more distracted user altering the free spreadsheet and, then, not being able to recompose it. The spreadsheet in Appendix E is blocked and hidden, constituting an exception. These appendices are found on the website: http://extras.springer.com.

Appendix A brings the solution of what was developed in Chap. 5; Appendix B brings a generic solution for what was proposed in Chap. 5; Appendix C contains the databases utilized in the development of five paragraphs of Chap. 6 and the exercises of Chaps. 6 and 8; Appendix D brings the solutions for what was developed in the text of Chap. 6 and provides the guidance for the exercises proposed in this chapter; and finally, Appendix E presents the solution for a challenge (exercise) proposed in Chap. 9.

Even though the language of logic is developed with all the strictness the subject demands, the exhibition of the book is pervaded by language abuse. The attentive reader will perceive them and be able to overcome them as he/she becomes acquainted with the text.

São Paulo, Brazil   Fábio Romeu de Carvalho
                                                                                        Jair Minoro Abe

Acknowledgements

We would like to express our very great appreciation to Prof. Dr. João Carlos Di Genio, Rector of Paulista University—UNIP, São Paulo, Brazil, for providing us with a distinct support as researchers at UNIP over the past years.

Contents

Chapter 1
Logic

1.1 Preliminary Concepts

In this chapter, an outline of logic will be performed, since the classical until the paraconsistent annotated, to provide the reader with an overview of this science. However, the basic tool for the decision support system that will be analyzed is in Chap. 2, paraconsistent annotated evidential logic E_t. This way, the more informed reader will be able to, with little loss for the comprehension of the decision method, go directly to Chap. 2.

Considering that this is a work of logic application in Engineering, some language abuses will be allowed, as well as some inaccuracies that do not correspond to this science (logic). This will be done to make it more intuitive and understandable by the readers who are unfamiliar with logic, enabling them to learn some elementary concepts more easily. It is evident that the subject will not be exhausted.

For the logical propositions, the quality of false or true is normally attributed, associating to it a **truth-value** "false" (**F** or **0**) or "true" (**V** or **1**).

In order to relate the sentences with each other, the connectives are used. The five most common ones are: negation ($\neg$), conjunction ($\wedge$), disjunction ($\vee$), implication ($\rightarrow$) and biconditional ($\leftrightarrow$).

The connective of **negation** ($\neg$) makes the negation of a sentence. For example, being **p** the sentence "John is mortal", its negation $\neg$**p** means "John is not mortal".

Valuation is the function $\mathcal{V} : \mathcal{F} \rightarrow \{1; 0\}$, that is, the function defined in the set of sentences $\mathcal{F}$ on the set of truth-values $\{1; 0\}$ or $\{V; F\}$. Thus, if $\mathbf{p} \in \mathcal{F}$ *is* true, $\mathcal{V}(\mathbf{p}) = 1$ and if p is false, $\mathcal{V}(\mathbf{p}) = 0$. Considering the classical principle of negation "If a sentence is true, its negation is false and vice versa", we have:

$$\mathcal{V}(\mathbf{p}) = 1 \Leftrightarrow \mathcal{V}(\neg\mathbf{p}) = 0 (\Leftrightarrow \text{ means } ''\text{if, and only if,}'').$$

The connective of the **conjunction**($\wedge$) enables to translate two predicates of the same being. For instance, the sentence $\mathbf{A} \equiv$ "John is retired **and** widowed", which

F. R. de Carvalho and J. M. Abe, *A Paraconsistent Decision-Making Method*,
Smart Innovation, Systems and Technologies 87,
https://doi.org/10.1007/978-3-319-74110-9_1

has the same logical meaning of the sentences $\mathbf{p} \equiv$ "John is retired" **and** $\mathbf{q} \equiv$ "John is widowed". We say that the first sentence is the **conjunction** of the two last ones and the representation $\mathbf{A} \equiv \mathbf{p} \wedge \mathbf{q}$ is used.

It is concluded that: $\mathcal{V}(\mathbf{p} \wedge \mathbf{q}) = 1 \Leftrightarrow \mathcal{V}(\mathbf{p}) = 1$ and $\mathcal{V}(\mathbf{q}) = 1.$
Where: $\mathcal{V}(\mathbf{p} \wedge \mathbf{q}) = 0 \Leftrightarrow \mathcal{V}(\mathbf{p}) = 0$ or $\mathcal{V}(\mathbf{q}) = 0.$

The connective of the **disjunction** ($\vee$) translates at least one out of two predicates of the same being. For instance, the sentence $\mathbf{A} \equiv$ "John is retired **or** widowed", which has the same logical meaning of the sentences $\mathbf{p} \equiv$ "John is retired" **or** $\mathbf{q} \equiv$ "John is widowed". We say that the first sentence is the **disjunction** of the two last ones and the representation $\mathbf{A} \equiv \mathbf{p} \vee \mathbf{q}$ is used.

It is concluded that: $\mathcal{V}(\mathbf{p} \vee \mathbf{q}) = 1 \Leftrightarrow \mathcal{V}(\mathbf{p}) = 1$ or $\mathcal{V}(\mathbf{q}) = 1.$
Where: $\mathcal{V}(\mathbf{p} \wedge \mathbf{q}) = 0 \Leftrightarrow \mathcal{V}(\mathbf{p}) = 0$ and $\mathcal{V}(\mathbf{q}) = 0.$

"If $\mathbf{p}$, then $\mathbf{q}$", is the same as "$\mathbf{p}$ implies $\mathbf{q}$" and this is a new sentence obtained from sentences $\mathbf{p}$ and $\mathbf{q}$. It will be represented by $\mathbf{p} \rightarrow \mathbf{q}$ and the connective ($\rightarrow$) that represents it is called **implication**; $\mathbf{p}$ receives the name of antecedent and $\mathbf{q}$, of consequent of the implication. It is verified that the antecedent of the implication is the sufficient condition for the consequent and it is the necessary condition for the former.

We have: $\mathcal{V}(\mathbf{p} \rightarrow \mathbf{q}) = 1 \Leftrightarrow \mathcal{V}(\mathbf{p}) = 0$ or $\mathcal{V}(\mathbf{q}) = 1.$
Where: $\mathcal{V}(\mathbf{p} \rightarrow \mathbf{q}) = 0 \Leftrightarrow \mathcal{V}(\mathbf{p}) = 1$ and $\mathcal{V}(\mathbf{q}) = 0.$

If $\mathbf{p}$ is a necessary and sufficient condition for $\mathbf{q}$, it is represented by $\mathbf{p} \rightarrow \mathbf{q}$ and the connective is called **biconditional**.

We have: $\mathcal{V}(\mathbf{p} \leftrightarrow \mathbf{q}) = 1 \Leftrightarrow \mathcal{V}(\mathbf{p}) = \mathcal{V}(\mathbf{q})$(both equal to 1 or both equal to 0).
Where: $\mathcal{V}(\mathbf{p} \leftrightarrow \mathbf{q}) = 0 \Leftrightarrow \mathcal{V}(\mathbf{p}) \neq \mathcal{V}(\mathbf{q}).$

The presented principles may be summarized by the denominated **truth tables**, represented in Table 1.1.

Observe that: $\mathcal{V}(\mathbf{p} \wedge \mathbf{q}) = \min\{\mathcal{V}(\mathbf{p}), \mathcal{V}(\mathbf{q})\}$ and
$\mathcal{V}(\mathbf{p} \vee \mathbf{q}) = \max\{\mathcal{V}(\mathbf{p}), \mathcal{V}(\mathbf{q})\},$

Table 1.1 Truth tables

p	q	¬p	p ∧ q	p ∨ q	p → q	p ↔ q
1	1	0	1	1	1	1
1	0	0	0	1	0	0
0	1	1	0	1	1	0
0	0	1	0	0	1	1

The simple sentences of the kind **p** or **q** are called atomic formulas; the composed ones, such as $\mathbf{A} = \neg\mathbf{p}$, $\mathbf{B} = \mathbf{p} \wedge \mathbf{q}$, $\mathbf{C} = \mathbf{p} \vee \mathbf{q}$, $\mathbf{D} = \mathbf{p} \to \mathbf{q}$ and $\mathbf{E} = \mathbf{p} \leftrightarrow \mathbf{q}$ are called complex formulas.

1.2 Classical Logic

In this paragraph, some important concepts related to the deductive part of classical logic will be presented, without concern with excessive strictness or richness of details.

(I) The first concept concerns the (inference) rule of ***modus ponens***, which enables, from the formulas **A** and **A** → **B**, to infer **B**, that is, if **A** and **A** → **B**, then **B**. This inference rule is of extreme importance in the study of logic and is represented as follows:

$$\frac{A, A \to B}{B}.$$

If **A** and **A** → **B** are true, **B** will also be true.

(II) Another concept that stands out is the concept of **demonstration** (or proof), defined as being a finite sequence of formulas $(\mathbf{A}_1, \mathbf{A}_2, ..., \mathbf{A}_n)$ $(n \geq 1)$, so that, whatever k is, $1 \leq k \leq n$:

(a) or $\mathbf{A}_k$ is an axiom;

(b) or $\mathbf{A}_k$ was obtained from $\mathbf{A}_i$, and $\mathbf{A}_j$, with $i < k$ and $j < k$, by the application of the *modus ponens* rule.

$$\frac{A_i, A_j}{A_k} \quad or \quad \frac{A_i, A_i \to A_k}{A_k}, \text{ where } A_i \to A_k \text{ is } A_j$$

(III) We say that a formula **A** of the language is a **theorem**, if a demonstration exists $(\mathbf{A}_1, \mathbf{A}_2, ..., \mathbf{A}_n)$ $(n \geq 1)$, so that $\mathbf{A}_n = \mathbf{A}$. The sequence $(\mathbf{A}_1, \mathbf{A}_2, ..., \mathbf{A}_n)$ is called demonstration if **A**. It is represented: $\vdash \mathbf{A}$.

(IV) Consider Γ a set of formulas. A **deduction**, from Γ, is any finite sequence of formulas $(\mathbf{A}_1, \mathbf{A}_2, ..., \mathbf{A}_n)$ $(n \geq 1)$, so that, for every k, $1 \leq k \leq n$:

(a) either $\mathbf{A}_k$ is an axiom;

(b) or $\mathbf{A}_k$ is an element of Γ;

(c) or $\mathbf{A}_k$ was obtained from $\mathbf{A}_i$, and $\mathbf{A}_j$, with $i < k$ and $j < k$, by the application of the *modus ponens* rule.

$$\frac{A_i, A_j}{A_k} \quad or \quad \frac{A_i, A_i \to A_k}{A_k}, \text{ where, evidently, } A_i \to A_k \text{ is } A_j.$$

The elements from Γ are called hypotheses (or premises).

(V) A formula **A** is said to be a **syntactic consequence** of a set of formulas Γ, if a deduction exists $(\mathbf{A}_1, \mathbf{A}_2,..., \mathbf{A}_n)$ $(n \geq 1)$, from Γ, so that $\mathbf{A}_n = \mathbf{A}$.
It is represented by $\Gamma \vdash \mathbf{A}$ or by $\mathbf{B}_1, \mathbf{B}_2,..., \mathbf{B}_m \vdash \mathbf{A}$ (without the set representation curly braces), if Γ is a finite set $\{\mathbf{B}_1, \mathbf{B}_2,..., \mathbf{B}_m\}$
Observe that, if $\Gamma = \varnothing$, $\Gamma \vdash A \Leftrightarrow \varnothing \vdash A \Leftrightarrow \vdash A$, that is, a theorem is a syntactic consequence of the empty set $\varnothing$.

(VI) A sentence (or formula) is called **tautology** (or **logically valid** sentence) when it is always true, whatever are the truth-values of its component sentences (or formulas). When it is always false, it is called a **contradiction**.

(VII) **Deduction Theorem**: Consider Γ a set of formulas and **A** and **B** two formulas.

(a) If $\Gamma, \mathbf{A} \vdash \mathbf{B}$, then $\Gamma \vdash \mathbf{A} \to \mathbf{B}$ (that is, if from Γ and **A** we deduce **B**, then from Γ we deduce $\mathbf{A} \to \mathbf{B}$).
In particular, we have:
(b) If, $\mathbf{A} \vdash \mathbf{B}$, then $\vdash \mathbf{A} \to \mathbf{B}$ (that is, if from **A** we deduce **B**, then $\mathbf{A} \to \mathbf{B}$ is a theorem).

An axiomatic of a calculus is constituted by its postulates (schemes of axioms and inference rules). Here, the axiomatic of Kleene [75] for the classical propositional calculus will be presented.
Consider any **A**, **B** and **C** formulas.

(a) Postulates of the implication:

(Al) $\mathbf{A} \to (\mathbf{B} \to \mathbf{A})$
(A2) $(\mathbf{A} \to \mathbf{B}) \to ((\mathbf{A} \to (\mathbf{B} \to \mathbf{C})) \to (\mathbf{B} \to \mathbf{C}))$
(A3) $\frac{A, A \to B}{B}$ (modus ponens rule)

(b) Conjunction axiom schemes:

(A4) $(\mathbf{A} \wedge \mathbf{B}) \to \mathbf{A}$
(A5) $(\mathbf{A} \wedge \mathbf{B}) \to \mathbf{B}$
(A6) $\mathbf{A} \to (\mathbf{B} \to (\mathbf{A} \wedge \mathbf{B}))$

(c) Disjunction axiom schemes:

(A7) $\mathbf{A} \to (\mathbf{A} \vee \mathbf{B})$
(A8) $\mathbf{B} \to (\mathbf{A} \vee \mathbf{B})$
(A9) $(\mathbf{A} \to \mathbf{C}) \to (\mathbf{B} \to \mathbf{C}) \to ((\mathbf{A} \vee \mathbf{B}) \to \mathbf{C})$

(d) Negation axiom schemes:

(A10) $\mathbf{A} \vee \neg\mathbf{A}$ (law of excluded middle)
(A11) $(\mathbf{A} \wedge \neg\mathbf{A}) \to \mathbf{B}$ or $\mathbf{A} \to (\neg\mathbf{A} \to \mathbf{B})$

(A12) $(\mathbf{A} \rightarrow \mathbf{B}) \rightarrow ((\mathbf{A} \rightarrow \neg\mathbf{B}) \rightarrow \neg\mathbf{A})$ (principle of the reduction to absurdity)

The postulates of groups (a), (b), (c) and (d) constitute the so-called **classical propositional logic**: L [→, ∧, ∨, ¬].
Notes:

(1) A proposition of the kind $\mathbf{A} \wedge \neg\mathbf{A}$ (which is false in the classical logic) is said to be a **contradiction** or **inconsistency**.
(2) The classical propositional logic is decidable by means of the truth-tables (Table 1.1) or matrices.
(3) The classical propositional calculus L [→, ∧, ∨, ¬] may be extended to the **classical predicate calculus**: L [→, ∧, ∨, ¬,∀, ∃] [82], (∀ is the universal quantifier and ∃, the existential quantifier).

(e) Axioms schemes and inference rule of the quantification:

(A14) $\forall x\, \mathbf{A}(x) \rightarrow \mathbf{A}(c)$

(A15) $\dfrac{\mathbf{A} \rightarrow \mathbf{B}(x)}{\mathbf{A} \rightarrow \forall x\mathbf{B}(x)}$

(A16) $\mathbf{A}(c) \rightarrow \exists x\mathbf{A}(x)$

(A17) $\dfrac{\mathbf{A}(x) \rightarrow \mathbf{B}}{\exists X\mathbf{A}(X) \rightarrow \mathbf{B}}$

with the usual restrictions.
The following equivalences are valid:

$$\exists x\, \mathbf{A}(x) \leftrightarrow \neg\forall x\neg\, \mathbf{A}(x) \quad \text{and} \quad \forall x\, \mathbf{A}(x) \leftrightarrow \neg\exists x\neg\, \mathbf{A}(x)$$
$$\neg\exists x\, \mathbf{A}(x) \leftrightarrow \forall x\neg\, \mathbf{A}(x) \quad \text{and} \quad \neg\forall x\, \mathbf{A}(x) \leftrightarrow \exists x\neg\, \mathbf{A}(x)$$

The classical predicate calculus is not decidable, except in some particular cases [66].

1.3 The Non-classical Logics

Still without much strictness, we may say that the non-classical logics compose two large groups:

(1) the ones that complement the scope of classical logic; and
(2) the ones that rival classical logic.

The logics belonging to the first category are called complementary of the classical and, as the name itself says, they complement aspects that classical logic is not able to express. They are based on the classical logic and broden its power of expression. They comprise, as an example:

- the epistemic logics (logics of belief, logics of knowledge, logics of doubt, logics of justification, logics of preference, logics of decision, logics of acceptance, logics of confirmation, logics of opinion, deontic logics, etc.);
- the traditional modal logic (T system, S4 system, S5 system, multi-modal systems, etc.);
- intentional logics;
- logics of action (logics of the imperative, logics of decision, etc.);
- logics for physical applications (logic of time (linear, non-linear, etc.), chronological logics, logics of space, Lésniewski logic, etc.);
- combinatory logics (related to λ-calculus);
- infinitary logics;
- conditional logics, etc.

In the second group are the logics that rival the classical logic (also called heterodox). They restrict or modify certain fundamental principles of traditional logic.

As it was commented at the beginning, besides the Fuzzy Logic, innumerous other heterodox systems have recently been cultivated, most of them mainly motivated by the advances experienced in this field of science, mostly by Artificial Intelligence:

- intuitionistic logics (Intuitionistic logic without negation, Griss logic, etc.). Such systems are well established (there is cultivated mathematics and they possess interesting philosophical characteristics);
- non-monotonic logics;
- linear logics;
- default logics;
- defeasible logics;
- abductive logics;
- multi-valued logics (or multipurpose logics: Lukasiewicz's logic, Post's logic, Gödel's logic, Kleene's logic, Bochvar's logic, etc.). Their studies are in advanced phase. In fact, there is a kind of constructed mathematics in these systems and they have philosophical importance, addressing, for example, the issue of future contingents;
- Rough set theory;
- paracomplete logics (that restrict the law of excluded middle);
- paraconsistent logics (that restrict the principle of non-contradiction: C_n systems, annotated logics, logics of paradox, discursive logics, dialectical logics, relevant logics, logics of inherent ambiguity, imaginary logics, etc.);
- non-alethic logics (logics that are simultaneously paracomplete and paraconsistent);
- non-reflexive logics (logics that restrict the principle of identity);
- self-referential logics;
- labeled logics, free logics, quantum logics, among others.

The non-classical systems have deep meaning, not only from the practical point of view, but also from the theoretical one, breaking a paradigm of human thought that has been ruling for more than two thousand years.

1.4   Paraconsistent Logic

Paraconsistent Logic had the Russian logician Vasiliev and the Polish logician Lukasiewicz as its pioneers. Both of them, in 1910, independently, published works that addressed the possibility of a logic that did not eliminate, *ab initio*, the contradictions. Nevertheless, these authors' works, concerning paraconsistency, were restricted to the traditional Aristotelian logic. Only in 1948 and 1954 the Polish logician Jaskowski and the Brazilian logician Newton da Costa, respectively and independently, constructed paraconsistent logic [11].

Jáskowski formalized a paraconsistent propositional calculus denominated Discursive (or Discussive) Propositional Calculus, whereas da Costa developed several paraconsistent logics containing all the common logical levels. Also, independently, Nelson, in 1959, investigated the constructive systems with strong negation closely related to the ideas of paraconsistency.

Consider $\mathcal{F}$ the set of all the sentences (or formulas) of the language $\mathcal{L}$ of a calculus (or logic) $\mathcal{C}$. Consider $\mathcal{T}$ a subset of $\mathcal{F}$. We say that $\mathcal{T}$ is a theory (of $\mathcal{C}$), if $\mathcal{T}$ is closed in relation to the notion of syntactic consequence of $\mathcal{C}$, that is,

$$\mathcal{T} = \{A : \mathcal{T} \vdash_{\mathcal{C}} A\},$$

that is, **A** is a syntactic consequence of $\mathcal{T}$ if and only if $A \in \mathcal{T}$ (Sometimes, we say that A is a "theorem" of $\mathcal{T}$, giving a broader sense (deductible from) to the word theorem). In this case, $\mathcal{C}$ is called underlying calculus or logic to the theory $\mathcal{T}$.

It is said that a theory $\mathcal{T}$, whose underlying logic is $\mathcal{C}$ and whose language is $\mathcal{L}$ is **inconsistent** if it contains at least one "theorem" **A** so that its negation ¬**A** is also a "theorem" of $\mathcal{T}$, that is, if at least one formula A of $\mathcal{F}$ exists, so that **A** and ¬**A** belong to $\mathcal{T}$ (they are theorems of $\mathcal{T}$). Otherwise, it is considered **consistent**.

A theory $\mathcal{T}$ is considered **trivial** when all the formulas of $\mathcal{F}$ are "theorems" of $\mathcal{T}$, that is, $\mathcal{T}$ is trivial if, and only if, $\mathcal{T} = \mathcal{F}$. Otherwise, it is considered **non-trivial**. A theory $\mathcal{T}$ is **paraconsistent** if it is **inconsistent and non-trivial** [33].

In classical logic, from **A** and ¬**A** it is possible to demonstrate any formula **B**. Therefore, if a classical theory has a contradiction, all the language formulas are theorems of this theory. That means that a contradiction trivializes any classical theory.

A logic (or calculus) is considered **paraconsistent** if it can be the underlying logic of paraconsistent theories (inconsistent, but non-trivial) [32]. Therefore, in the paraconsistent theories, there are formulas **A** so that, from **A** and ¬**A**, it is not possible to demonstrate any formula **B**, that is, there is always a formula **B** of $\mathcal{F}$ so that **B** is not a theorem of the theory.

In summary, a theory $\mathcal{T}$ is inconsistent if there is a formula **A** so that **A** and ¬**A** are both deductible from $\mathcal{T}$; otherwise, $\mathcal{T}$ is consistent. $\mathcal{T}$ is considered trivial if all the language formulas belong to $\mathcal{T}$, that is, if $\mathcal{T} = \mathcal{F}$; otherwise,

$\mathcal{T}$ is non-trivial. $\mathcal{T}$ os considered **paraconsistent** if it is **inconsistent** and **non-trivial**.

Paraconsistent logic (PL) was built o satisfy the following conditions: (a) in PL, in general, the principle of the non-contradiction must not be valid; (b) from one contradiction, it must not be possible to deduce every proposition.

Analogously, the same definition applied to proposition systems, set of information, etc. (taking into account, naturally, the set of its consequences).

In classical logic and in several logic categories, consistency plays an important role. As it was previously seen, in most of the usual logical systems, if a theory $\mathcal{T}$ is trivial, it is inconsistent and reciprocally.

A logic $\mathcal{C}$ is called **paraconsistent** if it can serve as a base (if it can be the underlying logic) for inconsistent theories, but non-trivial, that is, for paraconsistent theories.

A logic $\mathcal{C}$ is called **paracomplete** if it can be the underlying logic of theories in which the principle of the excluded third is infringed the following way: out of two contradictory propositions, one is true. Therefore, as it infringes, in this logic there may be two formulas **A** and ¬**A** both non-true.

Accurately, a logic is paracomplete if in it there are maximal non-trivial systems to which a certain formula and its negation do not belong.

Finally, a logic $\mathcal{C}$ is denominated **non-alethic** if it is paraconsistent and paracomplete.

In the positive part, da Costa's axiomatic (1993) for the paraconsistent logic is equal to Kleene's (1952) for the classical one. Therefore, they differ in the negation axioms. This way, items (a), (b) and (c), corresponding to the A1–A9 axioms, are identical to the ones of the classical, and the negation ones are the following:

(d') Negation axiom schemes:

    (A'10) **A** ∨ ¬**A**   (law of excluded middle)
    (A'11) ¬¬**A** → **A**   (property of double negation)
    (A'12) **B**° → ((**A** → **B**) → ((**A** → ¬**B**) → ¬**A**)) (principle of the reduction to absurdity)
    (A'13) **A**° ∧ **B**° → ((**A** → **B**)° ∧ (**A** ∧ **B**)°∧(**A** ∨ **B**)°)

where **B**° = $_{\text{def}}$¬ (**B** ∧ ¬**B**) is denominated **well-behaved formula**.

Note: From A13, it is concluded that "well-behaved property" is maintained in the implication, in the conjunction and in the disjunction, that is, formulas formed from well-behaved formulas are also well-behaved.

1.5 Paraconsistent Annotated Logic (PAL)

The paraconsistent annotated logics are a family of non-classical logics initially employed in logical programming by Subrahmanian [31]. Due to the obtained applications, a study of the foundations of the underlying logic of the investigated programming languages became convenient. It was verified that it was a para-consistent logic and that, in some cases, also contained characteristics of para-complete and non-alethic logic.

The first studies concerning the foundations of the PAL were conducted by da Costa, Vago, Subrahmanian, Abe and Akama [1, 8, 9, 30, 31]. In [1], the logic of predicates was studied, as well as theory of models, annotated theory of sets and some modal systems, establishing a systematic study of the foundations of the annotated logics pointed out in previous works. In particular, metatheorems of strong and weak completeness were obtained for a subclass of first order annotated logic, and a systematic study of the annotated theory of models was conducted, generalizing the majority of the standard results for the annotated systems.

Other applications of the annotated systems were initiated by Abe, around 1993, which, along with his disciples, implemented the paraconsistent programming language (*Paralog*), independently from Subrahmanian's results. Such ideas were applied in the construction of a prototype and in the specification of an architecture based on the PAL, which integrates several computational systems—planners, databases, vision systems, etc.—, in the construction of a manufacture cell and in the representation of knowledge by Frames, allowing to translate inconsistencies and exceptions.

da Silva Filho, another one of Abe's disciples, took interest for the application of PAL in digital circuits, obtaining the implementation of the logical ports Complement, And and Or. Such circuits enable "conflicting" signs implemented in its structure in a non-trivial manner. We believe the contribution of the paracon-sistent electric circuits is a pioneer in the electric circuit area, opening new investigations paths. Also in the researches about hardware, was the construction of the logical analyzer—para-analyzer—that enables to address concepts of uncer-tainty, inconsistency and paracompleteness. Logical controllers were also con-structed, based on the annotated logics—*Paracontrol*, logical simulators—*Parasim*, signal treatment—*Parassonic*.

As materialization of the discussed concepts, the first paraconsistent robot was built with the paraconsistent hardware: the Emmy robot. Another paraconsistent robot, built with the software based on PAL, was called Sofya; and several other subsequent prototypes were built: Amanda, Hephaestus, etc.

The annotated systems also embrace aspects of the concepts involved in non-monotonic thought, defeasible, default and deontic.

Versions of annotated logics also involve several aspects of the fuzzy logics. That may be seen under various angles. The annotated set theory encompasses *in totum* the fuzzy set theory [1]. Axiomatized versions of the fuzzy theory were obtained.

The hybrid controller *parafuzzy* was erected, which united characteristics of the annotated and fuzzy logics. Finally, algebraic aspects were also investigated by Abe and other interesting algebraizations have been studied by several authors.

1.6 Lattice Associated to the Paraconsistent Annotated Logic

The importance of language theory for the investigation of problems in science is widely known. Thus, a good solution for an inquiry may often depend deeply on the choice or on the discovery of a convenient language to represent the concepts involved, as well as to make reasonable inferences until reaching satisfactory solutions.

Concerning the applications, closely observing a set of information obtained regarding a certain theme, one may notice that such set encloses contradictory information that generate difficulties for description of vague concepts, as we already discussed in the introduction. In the case of contradiction, they are normally removed artificially, so as not to contaminate the dataset, or suffer a separate treatment, with extralogical devices.

However, the contradiction, most of the times, contains decisive information, as it is the encounter of two opposite truth-value threads. Thus, neglecting it is proceeding anachronistically. Consequently, we must seek languages that can coexist with such contradictions, without hindering the other information. Regarding the concept of uncertainty, we must think of a language that is able to capture and encircle the 'maximum' of 'information' of the concept. In order to obtain a language that may have these characteristics, we propose the procedure described as follows.

Our aim is to host the concepts of uncertainty, inconsistency and paracompleteness in their linguistic structure and think (mechanically) of their presence, with the hope that, with this drawing, language enables us to reach, capture and reflect better the nuances of reality in a different manner from the traditional ones. Thus, the intention is to be equipped with a proper language and deductive structure for a comprehension of problems under different angles and, perhaps, this way we may generate innovative solutions. For this task, the concepts of inconsistency and paracompleteness will be considered. They will be joined by the notions of truth and falsity. This way, four objects will be considered, which will be generically called annotation constants.

T   called inconsistent;
⊥   called paracomplete;
V   called true;
F   called false.

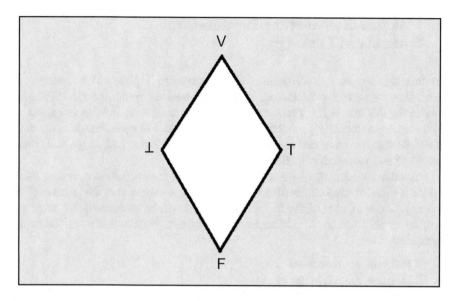

Fig. 1.1 Lattice 'four'

In the set of objects $\tau = \{T, V, F, \perp\}$ a mathematical structure will be defined: a **lattice** with operator $\tau = <|\tau|, \leq^*, \sim>$ characterized by the following Hasse diagram: (Fig. 1.1).

The operator $\sim :| \tau | \rightarrow | \tau |$ is defined this way:

- $\sim T = T$ (the 'negation' of an inconsistent proposition is inconsistent);
- $\sim V = F$ (the 'negation' of a 'true' proposition is 'false');
- $\sim F = V$ (the 'negation' of a 'false' proposition is 'true');
- $\sim \perp = \perp$ (the 'negation' of a 'paracomplete' proposition is 'paracomplete');

The operator $\sim$ will play the 'role' of the connective of the PAL negation, as it will be seen ahead.

The propositions of the PAL are of the p_μ type, where **p** is a proposition in the common sense and μ is an annotation constant.

Among several intuitive readings, p_μ may be read: 'I believe in proposition **p** with degree until μ or 'the favorable evidence expressed by the proposition is a maximum of μ'.

Suppose we have the proposition **p**: 'the front of the robot is free' and that there is information that allude to two situations:

1. 'the front of the robot is free' (that may be expressed in the PAL by p_V);
2. 'the front of the robot is not free' (that may be expressed in the PAL by p_F);

In a system based on the PAL, such situation may be represented by p_T: 'the front of the robot is free' constitutes an inconsistent state.

1.7   Axiomatization of the Paraconsistent Annotated Logic Qτ

In this section, an axiomatization of the annotated logics will be presented, extending the previous discussion, considering now an arbitrary lattice. The reference for this text is [1]. Technically, such logic constitutes in what is known as bi-sorted logic (intuitively speaking, there are two kinds of variables). It is worth to highlight that the annotated logics are paraconsistent logics and, in general, paracomplete and non-alethic, as it is exposed below.

Consider $\tau = <|\tau|, \leq^*, \sim>$ a finite lattice with fixed negation operator. Such lattice is called **truth-value lattice** and the operator $\sim$ constitutes the "meaning" of the negation symbol $\neg$ of the logical system that will be considered. Its language will be symbolized by $\mathcal{L} \ \tau$. Associated to lattice τ, we also have the following symbols:

- T indicates the maximum of τ;
- $\perp$ indicates the minimum of τ;
- sup indicates the supremum operation—regarding subsets of τ;
- inf indicates the infimum operation—regarding subsets of τ.

The language $\mathcal{L} \ \tau$ has the following primitive symbols:

1. Individual variables: a countable set of individual variables;

$$p, p_o, p_1, \ldots, q, q_o, q_1, \ldots, r, r_o, r_1, \ldots;$$
$$A, A_o, A_1, \ldots, B, B_o, B_1, \ldots, C, C_o, C_1, \ldots;$$

2. For each n, *n-ary* functional symbols; The 0-ary functional symbols are also called individual constants;
3. For each n, n-ary predicate symbols;
4. The equality symbol =;
5. Annotation constants (members of τ): $\mu, \lambda, \ldots$;
6. The symbols $\neg, \wedge, \vee, \rightarrow, \exists$ and $\forall$ of the connectives and of the quantifiers;
7. Auxiliary symbols: parenthesis and comma

The terms of language $\mathcal{L} \ \tau$ are defined in a usual manner. We utilize **a, b, c** and **d** —with or without indexes—as meta-variables for the terms.

Definition 1 [Formula] A **basic formula** is an expression such as $\mathbf{p}(a_1, \ldots, a_n)$, where **p** is an *n*-ary predicate symbol and $a_1, \ldots, a_n$ are terms of $\mathcal{L} \ \tau$. If $\mathbf{p}(a_1, \ldots, a_n)$ is a basic formula and $\mu \in \tau$ is an annotation constant, then $\mathbf{p_\mu}(a_1, \ldots, a_n)$ and $\mathbf{a} = \mathbf{b}$, where **a** and **b** are terms, are called **atomic formulas**.

The formulas have the following generalized inductive definition:

1. An atomic formula is a formula.
2. If **A** is a formula, then $\neg\mathbf{A}$ is a formula.

3. If **A** and **B** are formulas, then **A** ∧ **B**, **A** ∨ **B** and **A** → **B** are formulas.
4. If **A** is a formula and x is an individual variable, then (∃x)**A** and (∀x)**A** are formulas.
5. An expression of $\mathcal{L}$ τ constitutes a formula if, and only if, it is obtained applying one of the previous rules from 1 to 4 (maximal clause).

Formula ¬**A** is read "the *negation*—or *weak negation*—of **A**"; **A** ∧ **B**, "the *conjunction of A and* **B**"; **A** ∨ **B**, "disjunction of A and **B**"; **A** → **B**, "the *implication of* **B** *by* **A**"; (∃x)**A**, "the *instantiation of A by* x"; and (∀x)**A**, "the *generalization of A by x*".

Some symbols are introduced below, by definition:

Definition 2 [Equivalence and Strong Negation] Consider **A** and **B** any formulas of $\mathcal{L}$ τ. It is defined, then:

$$\mathbf{A} \leftrightarrow \mathbf{B} =_{def} (\mathbf{A} \to \mathbf{B}) \wedge (\mathbf{B} \to \mathbf{A}) \text{ and } \neg^* \mathbf{A} =_{def} \mathbf{A} \to ((\mathbf{A} \to \mathbf{A}) \wedge \neg(\mathbf{A} \to \mathbf{A})).$$

The symbol ¬* is called **strong negation**; therefore, ¬*A must be read the *strong negation of* **A**. The formula **A** ↔ **B** is read, as usually, the *equivalence of A and B*.

Definition 3 If **A** is a formula. Then:
$\neg^0\mathbf{A}$ indicates **A**; $\neg^1\mathbf{A}$ indicates ¬**A** and $\neg^k\mathbf{A}$ indicates $\neg(^{\neg k-1}\mathbf{A})$, ($k \in N$, $k > 0$).
Also, if $\mu \in |\tau|$, it is established that:
$\sim^0\mu$ indicates μ; $\sim^1\mu$ indicates $\sim\mu$ and $\sim^k\mu$ indicates $\sim(\sim^{k-1}\mu)$, ($k \in N$, $k > 0$).

Definition 4 [Literal] If $\mathbf{p}_\mu(a_1, \ldots, a_n)$ is an atomic formula. Any formula such as $\neg^k\mathbf{p}_\mu(a,\ldots\ldots, a)$ ($k > 0$) is called a **hyper-literal formula** or, simply, **literal**. The other formulas are called **complex formulas**. A semantics for the languages $\mathcal{L}$ τ is described now.

Definition 5 [Structure] A **structure** $\mathcal{E}$ for a language $\mathcal{L}$ τ consists of the following objects:

1. A non-empty set $|\mathcal{E}|$ denominated the **universe** of $\mathcal{E}$. The elements of $|\mathcal{E}|$ are called **individuals** of $\mathcal{E}$.
2. For each functional *n*-ary symbol f of $\mathcal{L}$ τ, an *n*-ary operation $f_\mathcal{E}$ of $|\mathcal{E}|$ in $|\mathcal{E}|$—in particular, for each individual constant e of $\mathcal{L}$ τ, e_E is an individual of $\mathcal{E}$.
3. For each predicate symbol **p** of weight n of $\mathcal{L}$ τ, a function $\mathbf{p}_E$: $|\mathcal{E}|^n \to |\tau|$.

Consider $\mathcal{E}$ a structure for $\mathcal{L}$ τ. The language-diagram $\mathcal{L}$ τ ($\mathcal{E}$) *is* obtained the usual manner. Given a term free of variable **a** of $\mathcal{L}$ τ ($\mathcal{E}$), it is also defined, commonly, the individual $\mathcal{E}(\mathbf{a})$ of $\mathcal{E}$. i and j are utilized as meta-variables to denote names.

The truth-value $\mathcal{E}(\mathbf{A})$ of the closed formula **A** of $\mathcal{L}$ τ ($\mathcal{E}$) is defined now. The definition is obtained by induction on the length of **A**. By language abuse, the same symbols are utilized for meta-variables of terms of the language-diagram.

Definition 6 Consider A a closed formula and L an interpretation for Lx.

1. If **A** is atomic of the form $\mathbf{p}_\mu$ $(a_1,\ldots, a_n)$, then:

   $\mathcal{E}$ $(A) = 1$ if, and only if, $\mathbf{p}_E$ ($\mathcal{E}$ $(a_1),\ldots, \mathcal{E}$ $(a_n)) \geq \mu$;
   $\mathcal{E}$ $(A) = 0$ if, and only if, it is not the case that $\mathbf{p}_E$ ($\mathcal{E}$ $(a_1),\ldots, \mathcal{E}$ $(a_n)) \geq \mu$;

2. If **A** is atomic of the form $\mathbf{a} = \mathbf{b}$, then:

   $\mathcal{E}$ $(A) = 1$ if, and only if, $\mathcal{E}$ $(\boldsymbol{a}) = \mathcal{E}$ $(\boldsymbol{b})$;
   $\mathcal{E}$ $(A) = 0$ if, and only if, $\mathcal{E}$ $(\boldsymbol{a}) \neq \mathcal{E}$ $(\boldsymbol{b})$;

3. If **A** is of the form $\neg^k(\mathbf{p}_\mu(a_1,\ldots,a_n))$ $(k \geq 1)$, then:
   $\mathcal{E}$ $(A) = \mathcal{E}$ $(\neg^{k-1}(\mathbf{p}_{\sim\mu}(a_1,\ldots,a_n)))$.

4. If **A** and **B** are any closed formulas, then:

   $\mathcal{E}$ $(\mathbf{A} \wedge \mathbf{B}) = 1$, if, and only if, $\mathcal{E}$ $(A) = \mathcal{E}$ $(B) = 1$;
   $\mathcal{E}$ $(\mathbf{A} \vee \mathbf{B}) = 1$ if, and only if, $\mathcal{E}$ $(A) = 1$ *or* $\mathcal{E}$ $(B) = 1$;
   $\mathcal{E}$ $(\mathbf{A} \rightarrow \mathbf{B}) = 1$ if, and only if, $\mathcal{E}$ $(A) = 0$ *or* $\mathcal{E}(B) = 1$;

5. If **A** is a complex closed formula, then:
   $\mathcal{E}(\neg\mathbf{A}) = 1 - \mathcal{E}(\mathbf{A})$

6. If **A** is of the form $(\exists x)\mathbf{B}$, then:
   $\mathcal{E}$ $(A) =$ if, and only if, $\mathcal{E}$ $(B_x[i]) = 1$ for some i in $\mathcal{L}$ τ $(\mathcal{E})$.

7. If **A** is of the form $(\forall x)\mathbf{B}$, then:
   $\mathcal{E}$ $(A) =$ if, and only if, $\mathcal{E}(B_x[i]) = 1$ for the every i in $\mathcal{L}$ τ $(\mathcal{E})$.

Theorem 1 *Consider A, B, C any formulas of* $\boldsymbol{Q}\tau$. *The connectives* $\rightarrow$, $\wedge$, $\vee$, $\neg^*$, *together with the quantifiers* $\forall$ *and* $\exists$, *have all the classical properties of implication, disjunction, conjunction and negation, as well as of the classical quantifiers* $\forall$ *and* $\exists$, *respectively. For example, we have that:*

1. $\vdash \neg * \exists x\boldsymbol{B} \vee \boldsymbol{C} \leftrightarrow \exists x(\boldsymbol{B} \vee \boldsymbol{C})$;
2. $\vdash \neg * \exists x\boldsymbol{B} \vee \exists x\boldsymbol{C} \leftrightarrow \exists x(\boldsymbol{B} \vee \boldsymbol{C})$;
3. $\vdash \neg * \forall x\boldsymbol{A} \leftrightarrow \neg * \exists x \neg * \boldsymbol{A}$;
4. $\vdash \neg * \exists x\boldsymbol{A} \leftrightarrow \neg * \forall x \neg * \boldsymbol{A}$;

The system of postulates—axiom schemes and inference rules—for $\mathbf{Q}\tau$, which is presented below, will be denominated $\mathbf{A}\tau$.

In the positive part, the axiomatic of PAL is equal to the one of the classical (items (a), (b) and (c)) added of Peirce's Law (item A″13). The negation axioms are A″10, A″11 and A″12, below [30, 37]:

(d″) Negation axiom schemes

Being **A**, **B**, **C** any formulas, **F** and **G** complex formulas, **p** a propositional variable and μ, μ_j $1 \leq j \leq n$, annotation constants, x, $x_1,\ldots, x_n$, $y_1,\ldots, y_n$ individual variables, we have:

(A″10) $\mathbf{F} \vee \neg\mathbf{F}$
(A″11) $\mathbf{F} \to (\neg\mathbf{F} \to \mathbf{A})$
(A″12) $(\mathbf{F} \to \mathbf{G}) \to ((\mathbf{F} \to \neg\mathbf{G}) \to \neg\mathbf{F})$
(A″13) $((\mathbf{A} \to \mathbf{B}) \to \mathbf{A}) \to \mathbf{A}$ (Peirce's Law)

(e″) Axioms schemes and inference rule of the quantification:

(A″14) $\forall x\mathbf{A}(x) \to \mathbf{A}(c)$
(A″15) $\dfrac{\mathbf{A} \to \mathbf{B}(X)}{\mathbf{A} \to \forall X\mathbf{B}(x)}$
(A″16) $\mathbf{A}(c) \to \exists x\mathbf{A}(x)$
(A″17) $\dfrac{\mathbf{A}(X) \to \mathbf{B}}{\exists X\mathbf{A}(X) \to \mathbf{B}}$

(f″) Axiom schemes specific of the PAL:

(A″18) $\mathbf{p}_{\perp}$
(A″19) $(\neg^k\mathbf{p}_\mu) \leftrightarrow (\neg^{k-1}\mathbf{p}_{\sim\mu})\ k \geq 1$
(A″20) $\mathbf{p}_\mu \to \mathbf{p}_\lambda$, where $\mu \geq \lambda$
(A″21) $\mathbf{p}_{\mu 1} \wedge \mathbf{p}_{\mu 2} \wedge \ldots \wedge \mathbf{p}_{\mu n} \to \mathbf{p}_\mu$, where $\mu = \sup \mu_{j},\ j = 1, 2, \ldots, n$
(A″22) $x = x$
(A″23) $x_1 = y_1 \to \ldots \to x_n = y_n \to \mathbf{f}(x_1, \ldots x_n) = \mathbf{f}(y_1, \ldots, y_n)$
(A″24) $x_1 = y_1 \to \ldots \to x_n = y_n \to \mathbf{P}_\mu(x_1, \ldots x_n) = \mathbf{P}_\mu(y_1, \ldots, y_n)$ with   the
       usual restrictions.

Theorem 2 *Qτ is paraconsistent if, and only if, #τ $\geq 2^1$. (#τ = cardinality of τ).*

Theorem 3 *If Qτ is paracomplete, then #τ ≥ 2. If #τ ≥ 2, there are Qτ systems that are paracomplete and there are Qτ that are not paracomplete.*

Theorem 4 *If Qτ is non-alethic, then #τ ≥ 2. If #τ ≥ 2, there are Qτ systems that are non-alethic and Qτ systems that are not non-alethic.*
Consequently, we see that the **Qτ** systems are, in general, paraconsistent, paracomplete and non-alethic.

Theorem 5 *The calculus Qτ is non-trivial.*
In [1] soundness and completeness theorems were demonstrated for the **Qτ** calculus when the lattice is finite[2]. Besides that, J.M. Abe showed how the standard model theory may be extended to the annotated logics of 1st order. In this same reference, it is evidenced that the annotated set theory is extraordinarily strong, involving, as this specific case, the Fuzzy set theory.
As a consequence, Annotated Mathematics, which involves Fuzzy Mathematics, seems to be of high relevance; just remember the applications made in Computing

---

[1] The symbol # indicates the cardinal number of τ.
[2] When the lattice is endless, due to scheme (τ_4), we fall into an infinitary logic, which still needs to be investigated.

Science and the meaning of the point of view of the fuzzy logics and mathematics applications.

Paraconsistent annotated logic, still very young, discovered at the dusk of the past century, is one of the greatest achievements in the field of the non-classical logics of the latest times. Its composition, as bi-sorted logic, in which one of the variables possesses a mathematical structure, has produced incredible results regarding computability and electronic implementations. It constitutes a new alternative logic, extremely interesting, capable of manipulating concepts such as the ones of uncertainty, inconsistency and paracompleteness inside it, with very natural computational implementations and electronics.

We believe APL has very broad horizons, with enormous application potential and also as foundation to clarify the common ground of several non-classical logics. Perhaps, some day, it might even rival the fuzzy logic regarding the applications.

Chapter 2
Paraconsistent Annotated Evidential Logic Eτ

In this chapter, we detail the necessary theoretical basis for the proposed model (which will be referred some times as "paraconsistent model") for the investigations, as well as for the applications presented in this work.

2.1 General Aspects

The Paraconsistent Annotated Evidential Logic Eτ is presented, which is the theoretical basis for the model to be discussed. Such choice is because this logic enables us, as already mentioned, to manipulate inaccurate, inconsistent and paracomplete data.

In the logic, Eτ, each proposition $\mathbf{p}$, is associated to an annotation constant constituted of a pair $(a; b)$, representing as follows: $\mathbf{p}_{(a;\, b)} \cdot a$ and b vary in the real closed interval [0, 1]. Therefore, the pair $(a; b)$ belongs to the cartesian product [0, 1] × [0, 1]. Intuitively, a represents the degree of favorable evidence (or degree of belief) expressed in $\mathbf{p}$, and b, the degree of contrary evidence (or degree of disbelief) expressed in $\mathbf{p}$. The pair $(a; b)$ is called annotation constant or, simply, annotation and may be represented by μ. Thus, it is written: $\mu = (a; b)$. The atomic propositions of the logic Eτ are the kind $\mathbf{p}_{\mu}$ or $\mathbf{p}_{(a;b)}$.

This way, some extreme situations may be highlighted, which correspond to the so-called extreme (or cardinal) states.

$\mathbf{p}_{(1;0)}$ represents maximum favorable evidence and no contrary evidence in $\mathbf{p}$; it is said that proposition $\mathbf{p}$ is **true** (V) and that the pair **(1, 0)** translates the state of **truth** (V).

$\mathbf{p}_{(0;1)}$ represents no favorable evidence and maximum contrary evidence in $\mathbf{p}$; it is said that proposition $\mathbf{p}$ is **false (F)** and that the pair (0; 1) translates the state of falsity (F).

© Springer International Publishing AG, part of Springer Nature 2018
F. R. de Carvalho and J. M. Abe, *A Paraconsistent Decision-Making Method*,
Smart Innovation, Systems and Technologies 87,
https://doi.org/10.1007/978-3-319-74110-9_2

$\mathbf{p}_{(1;1)}$ represents maximum favorable evidence and maximum contrary evidence in $\mathbf{p}$; it is said that proposition $\mathbf{p}$ is **inconsistent** ($\top$) and that the pair (1; 1) translates the state of **inconsistency** ($\top$).

$\mathbf{p}_{(0;0)}$ represents no favorable evidence and no contrary evidence in $\mathbf{p}$; it is said that proposition $\mathbf{p}$ is **paracomplete** ($\bot$) and that the pair (0; 0) translates the state of paracompleteness ($\bot$).

$\mathbf{p}_{(0.5;0.5)}$ may be read as an indefinite proposition (favorable evidence and contrary evidence equal to 0.5).

We observe that the paracomplete concept is the dual of the inconsistency concept.

Example 1 Consider the proposition $\mathbf{p} \equiv$ "Carnaby Street is adequate for the installation of the new enterprise". Then, we have:

$\mathbf{p}_{(1.0;0.0)}$ may be read as "Carnaby Street is adequate for the installation of the new enterprise", with total favorable evidence and null contrary evidence. Intuitively, it is a true proposition.

$\mathbf{p}_{(0.0;1.0)}$ may be read as "Carnaby Street is adequate for the installation of the new enterprise", with null favorable evidence and total contrary evidence. Intuitively, it is a false proposition.

$\mathbf{p}_{(1.0;1.0)}$ may be read as "Carnaby Street is adequate for the installation of the new enterprise", with total favorable evidence and also total contrary evidence. Intuitively, it is a contradictory proposition.

$\mathbf{p}_{(0.0;0.0)}$ may be read as "Carnaby Street is adequate for the installation of the new enterprise", with null favorable evidence and also null contrary evidence. Intuitively, it is a paracomplete proposition.

$\mathbf{p}_{(0.5;0.5)}$ may be read as "Carnaby Street is adequate for the installation of the new enterprise", with favorable evidence equal to 0.5 and contrary evidence also equal to 0.5. Intuitively, we have there an indefinition.

Based on the extreme (or cardinal) states, and by the use of the properties of the real numbers, carefully, a mathematical structure will be constructed with the purpose of materializing the ideas of how we want to handle mechanically the concepts of uncertainty, contradiction and paracompleteness, among others. Such mechanism will encompass, naturally, somehow, the true and false states addressed within the scope of classical logic, with all their consequences.

Example 2 Consider the proposition $\mathbf{p} \equiv$ "The student passed the exam". Then, we have:

If it is annotated with (1.0; 0.0), the intuitive reading will be "The student passed the exam", with total favorable evidence (= there is total evidence that the student passed the exam).

If it is annotated with (0.0; 1.0), the intuitive reading will be "The student passed the exam", with total contrary evidence (= there is total evidence that the student failed the exam).

If it is annotated with (1.0; 1.0), the intuitive reading will be "The student passed the exam", with totally inconsistent evidence. That may occur if the student did not study enough and at the same time a friend says he/she saw him confident after the exam.

If it is annotated with (0.0; 0.0), the intuitive reading will be "The student passed the exam", with a total absence of evidence, neither favorable, nor contrary.

Example 3 Consider **p** the proposition "The most popular song by Johnny Mathis is Misty" and **q** the proposition "Johnny Mathis will go down in history of popular music". Then, the conjunction $\mathbf{p}_{(1.0;\ 0.0)} \wedge \mathbf{q}_{(0.9;\ 0.1)}$ is read as "The most popular song by Johnny Mathis is Misty", with total favorable evidence and no contrary evidence and "Johnny Mathis will go down in history of popular music", with 90% favorable evidence and 10% contrary evidence. This corresponds to "It is certain that the most popular song by Johnny Mathis is Misty" and it is practically certain that he will go down in the history of popular music.

Example 4 Consider **p** the proposition "The robot's route is to the right". It is read, then, the implication $\mathbf{p}_{(0.7;\ 0.6)} \rightarrow \mathbf{p}_{(0.5;0.4)}$ as "The robot's route is to the right", with 70% favorable evidence and 60% contrary evidence, **entails** "The robot's route is to the right", 50% favorable evidence and 40% contrary evidence.

Example 5 Consider **p** the proposition "The patient has the flu". Then, the equivalence $\mathbf{p}_{(0.7;0.2)} \rightarrow \mathbf{p}_{(0.2;0.7)}$ is read as "The patient has the flu", with 70% favorable evidence and 20% contrary evidence, which is equivalent to saying that it is not the case that "The patient has the flu", with 20% favorable evidence and 70% contrary evidence or "The patient does not have the flu", with 20% favorable evidence and 70% contrary evidence.

2.2 Lattice of the Annotation Constants

Consider $|\tau| = [0, 1] \times [0, 1]$, that is, $|\tau|$ is the cartesian product of the real unit closed interval [0, 1] for itself, which may also be represented by $[0, 1]^2$. In the set $|\tau| \times |\tau|$ a total order relation $\leq^*$ is defined as follows:

$$((a_1; b_1), (a_2; b_2)) \in\ \leq^* \quad \text{or} \quad (a_1; b_1) \leq^* (a_2; b_2) \Leftrightarrow a_1 \leq a_2 \quad \text{and} \quad b_2 \leq b_1$$

being $\leq$ the usual order relation of the real numbers.

The structure $\tau =\ <|\tau|, \leq^* >$ is a fixed lattice, called lattice τ of the annotations. The pair (0; 0) is represented by $\perp$; the pair (1; 1) is represented by $\top$. A representation of this lattice may be made by the so-called generalized Hasse diagram (Fig. 2.1).

Examples:

(a) $(0.6; 0.4) \leq^* (0.8; 0.3)$;
(b) $(0.5; 0.5) \leq^* (0.7; 0.5)$;
(c) $(0.8; 0.6) \leq^* (0.8; 0.5)$;

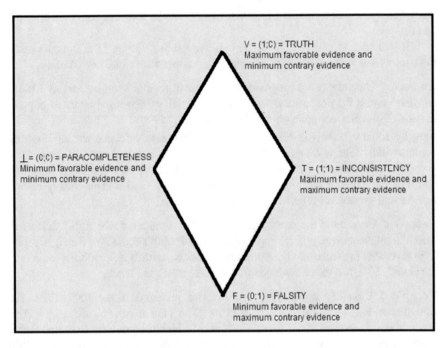

Fig. 2.1 Representation of lattice τ of the annotations by the generalized Hasse diagram

(d) $(0.8; 0,8) \leq^* (0.8; 0.8)$;
(e) $(a; b) \leq^* (1.0; 0.0)$ for any $0 \leq a, b \leq 1$;
(f) $(0.0; 1.0) \leq^* (a; b)$ for any $0 \leq a, b \leq 1$;

From (e) and (f) we obtain $(0.0; 1.0) \leq^* (a; b) \leq^* (1.0; 0.0)$ for any $0 \leq a, b \leq 1$.

Counterexamples

(a) it is false that $(0.9; 0.7) \leq^* (0.8; 0.6)$;
(b) it is false that $(0.4; 0.6) \leq^* (0.8; 0.7)$;
(c) it is false that $(0.9; 0.4) \leq^* (0.8; 0.6)$.

Properties

(a) $\forall a, b \in \tau, (a; b) \leq^* (a; b)$ (reflexivity);
(b) $\forall a_1, b_1, a_2, b_2 \in \tau, (a_1, b_1) \leq^* (a_2, b_2)$ and $(a_2, b_2) \leq^* (a_1, b_1)$ imply $(a_1, b_1) = (a_2, b_2)$ (anti–symmetry);
(c) $\forall a_1, b_1, a_2, b_2, a_3, b_3 \in \tau, (a_1, b_1) \leq^* (a_2, b_2)$ and $(a_2, b_2) \leq^* (a_3, b_3)$ imply $(a_1, b_1) \leq^* (a_3, b_3)$ (transitivity);
Properties (a)–(c) enable us to say that $\leq^*$ is a relation of order in $|\tau| \times |\tau|$.

(d) $\forall \, a_1, b_1, a_2, b_2, \in \tau$, there is the supremum of $\{(a_1, b_1); (a_2, b_2)\}$ indicated by
      sup $\{(a_1, b_1), (a_2, b_2)\} = (\max\{a_1, a_2\}; \min\{b_1, b_2\})$;
(e) $\forall \, a_1, b_1, a_2, b_2, \in \tau$, there is the infimum $\{(a_1, b_1); (a_2, b_2)\}$ indicated by inf
      $\{(a_1, b_1), (a_2, b_2)\} = (\min\{a_1, a_2\}; \max\{b_1, b_2\})$;
(f) $\forall a, b, \in \tau; (0; 1) \leq * (a; b) \leq * (1; 0)$. Therefore, inf $\tau = (0; 1)$ and sup $\tau = (1; 0)$.

The previous properties [from (a) to (f)] enable us to say that the set $[0, 1]^2$ with
the order $\leq *$ constitutes a lattice.

Exercise. Verify each one of the properties above.

Graphic interpretation
Regarding Fig. 2.2, we have that all the annotation constants $(a; b)$ of the shaded
region, including the edges, are such that $(a; b) \leq * (a_2; b_1)$ and, also, $(a_1; b_2) \leq *$
$(a; b)$. Only for $(a_2; b_1)$ and $(a_1; b_2)$, these properties are valid about the annotations
$(a; b)$ of the shaded region. $(a_2; b_1)$ is the supremum of the shaded region and
$(a_1; b_2)$ is it's infimum.

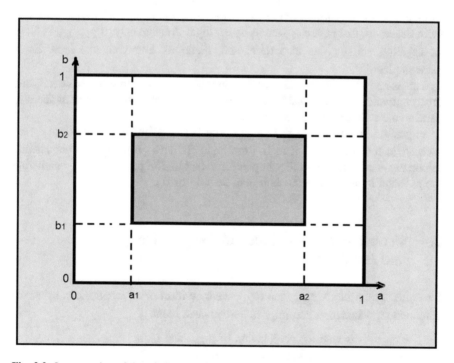

Fig. 2.2 Interpretation of the relation of order $\leq *$ of the lattice τ

2.3 Negation Connective

Operator $\sim: |\tau| \to |\tau|$, is defined according to [37], the following way:

$\sim (a; b) = (b; a)$. It has the "meaning" of the logical negation of Eτ.

Concerning this operator, the following comment is appropriate. Imagine the following sentence: "The Brazilian team will be the Olympic champion", with favorable evidence of 0.9 (or degree of belief = 0.9) and with contrary evidence of 0.2 (or degree of disbelief = 0.2). Its negation is "The Brazilian team will be the Olympic champion", with favorable evidence of 0.2 (or degree of belief = 0.2) and with contrary evidence of 0.9 (or degree of disbelief = 0.9). Therefore,

$$\neg \mathbf{P}_{(0.9;0.2)} \leftrightarrow \mathbf{P}_{(0.2;0.9)} \leftrightarrow \mathbf{P}_{(\sim (0.9;0.2))} \text{ or generalizing}$$
$$\neg \mathbf{P}_{(a;b)} \leftrightarrow \mathbf{P}_{(b;a)} \leftrightarrow P_{(\sim (a;b))}$$

Note that $\neg \mathbf{P}_{(0.5;0.5)} \leftrightarrow \mathbf{P}_{(0.5;0.5)}$, that is, $\mathbf{P}_{(0.5;0.5)}$ is equivalent to its negation $\neg \mathbf{P}_{(0.5;0.5)}$. Therefore, if $\mathbf{P}_{(0.5;0.5)}$ is true, its negation $\neg \mathbf{P}_{(0.5;0.5)}$ is also true, that is, a formula and its negation are both true. Therefore, the logic Eτ intuitively accepts inconsistencies, that is, it is a paraconsistent logic. Analogously, if $\mathbf{P}_{(0.5;0.5)}$ is false, its negation $\neg \mathbf{P}_{(0.5;0.5)}$ is also false, and, therefore, intuitively the logic Eτ is paracomplete.

This is an interesting property of the logic Eτ: to present true contradictions and false contradictions. This enables us to say that the logic Eτ paraconsistent and also paracomplete. Therefore, Eτ is non-alethic.

In general, we have $\neg \mathbf{P}_{(a;b)} \leftrightarrow \mathbf{P}_{(b;a)}$. The fact that the logical negation is "absorbed" in the annotation, it makes the logic Eτ have properties of fundamental importance and extreme fertility in paraconsistent logical programming, facilitating the physical implementations, as it may be seen in [1].

2.4 Connectives of Conjunction, Disjunction and Implication

Given the propositions $\mathbf{p}_{(a;b)}$ and $\mathbf{q}_{(c;d)}$ we may form other propositions using the conjunction, disjunction and implication between them:

$\mathbf{p}_{(a;b)} \wedge \mathbf{q}_{(c;d)}$ is read the conjunction of $\mathbf{p}_{(a;b)}$ and $\mathbf{q}_{(c;d)}$
$\mathbf{p}_{(a;b)} \vee \mathbf{q}_{(c;d)}$ is read the disjunction of $\mathbf{p}_{(a;b)}$ and $\mathbf{q}_{(c;d)}$
$\mathbf{p}_{(a;b)} \to \mathbf{q}_{(c;d)}$ is read the implication of $\mathbf{q}_{(c;d)}$ by $\mathbf{p}_{(a;b)}$

The connective of biconditional is introduced in the usual manner:

$$\mathbf{p}_{(a;b)} \leftrightarrow \mathbf{q}_{(c;d)} = \textbf{Def. } \left(\mathbf{p}_{(a;b)} \rightarrow \mathbf{q}_{(c;d)}\right) \wedge \left(\mathbf{q}_{(c;d)} \rightarrow \mathbf{p}_{(a;b)}\right)$$

is read $\mathbf{p}_{(a;b)}$ is equivalent to $\mathbf{q}_{(c;d)}$.

Example Suppose that the proposition $\mathbf{p}_{(0.6;0.4)}$ is "It will rain tomorrow", with 60% favorable evidence and 40% contrary evidence, and that the proposition $\mathbf{q}_{(0.3;0.6)}$ is "It will be cold tonight", with 30% favorable evidence and 60% contrary evidence.

The proposition $\mathbf{p}_{(0.6;0.4)} \wedge \mathbf{q}_{(0.3;0.6)}$ must be read as "It will rain tomorrow", with 60% favorable evidence and 40% contrary evidence, and "It will be cold tonight", with 30% favorable evidence and 60% contrary evidence.

The proposition $\mathbf{p}_{(0.6;0.4)} \vee \mathbf{q}_{(0.3;0.6)}$ must be read as "It will rain tomorrow", with 60% favorable evidence and 40% contrary evidence, **or** "It will be cold tonight", with 30% favorable evidence and 60% contrary evidence.

The proposition $\mathbf{p}_{(0.6;0.4)} \rightarrow \mathbf{q}_{(0.3;0.6)}$ must be read as "It will rain tomorrow", with 60% favorable evidence and 40% contrary evidence **entails that** "It will be cold tonight", with 30% favorable evidence and 60% contrary evidence.

We may introduce the connective of biconditional the usual manner:

$$\mathbf{p}_{(a;b)} \leftrightarrow \mathbf{q}_{(c;d)} - \text{is read } \mathbf{p}_{(a;b)} \text{ equivalent to } \mathbf{q}_{(c;d)}.$$

Thus, the proposition $\mathbf{p}_{(a;b)} \leftrightarrow \mathbf{q}_{(c;d)}$ must be read as "It will rain tomorrow", with 60% favorable evidence and 40% contrary evidence corresponds to "It will be cold tonight", with 30% favorable evidence and 60% contrary evidence.

2.5 Lattice τ

The set of the annotation constants $(a; b)$ may be represented in the cartesian coordinate system by the unit square $[0, 1] \times [0, 1]$, called unit square of the cartesian plane (USCP), which represents the lattice τ. A point $\mathbf{X} = (a; b)$ of this square represents the generic proposition $\mathbf{p}_{(a; b)}$.

To enable us to deal mechanically with the concepts of inaccuracy, inconsistency and paracompleteness, as well as with the concepts of truth and falsity, some definitions will be introduced.

Observe that the ends of the line segment CD (Fig. 2.3) are points that translate situations of perfect definition (the truth of falsity). For that reason, segment CD is called **perfectly defined line (or segment)** (PDL). The equation of this line is $a + b - 1 = 0$.

As you move away from line **CD** in the direction of point **A** or point **B**, the uncertainties gradually increase. When there is a movement in the direction from the PDL to **B**, both evidences, favorable and contrary, increase, tending to 1. Therefore, we tend to big evidences, favorable and contrary (next to 1), which represents a situation of uncertainty called inconsistency (or contradiction).

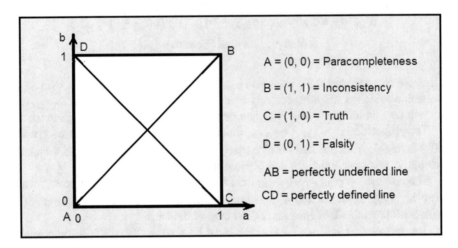

Fig. 2.3 Unit square of the cartesian plane (SUCP)

Accordingly, when there is a movement in the direction from the PDL to **A**, the evidences, favorable and contrary, decrease, tending to 0. In this case, we tend to small degrees of evidences, favorable and contrary (next to 0), which represents a situation of uncertainty called paracompleteness.

In view of the exposed, it is very reasonable to define the degree of uncertainty of the annotation $(a; b)$ as being

$$G(a; b) = a + b - 1,$$

which translates the distance from point $X = (a; b)$ to line **CD**, multiplied by $\sqrt{2}$ and affected by the signal + or − (here, metric distance was considered, as in analytical geometry).

Observe that $-1 \leq G \leq 1$.

Example G (0.8; 0.9) = 0.8 + 0.9 − 1 = 0.7. Therefore, the degree of uncertainty of the annotation constant (0.8; 0.9) is 0.7; it is a positive and relatively high value, pointing out to a high degree of inconsistency (or of contradiction). On the other hand, G (0.2;0.1) = 0.2 + 0.1 − 1 = −0.7; is a negative and relatively high value (in module), pointing out to a high degree of paracompleteness.

Each value of the degree of uncertainty belonging to the open interval]−1, 1 [defines a segment, parallel to PDL. If 0 < G < 1, this segment is called **inconsistency limit line** (RS, in Fig. 2.4); if −1 < G < 0, **paracompleteness limit line** (MN, in Fig. 2.4); if G = 0, it is the perfectly defined line (PDL); if G = −1 or 1, we have point **A** or point **B**, which are the extreme states of paracompleteness and of inconsistency, respectively.

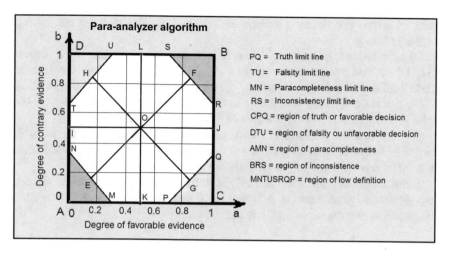

Fig. 2.4 QUPC, the four extreme regions and the limit lines

Line **AB** of the SUCP is called **perfectly undefined line (or segment)** (PUL). In fact, in all the points of the segment **AB**, the values of the favorable evidence (or degree of belief) and of the contrary evidence (or degree of disbelief) are equal (**a = b**). Therefore, they are points in which the mentioned values may be both small, characterizing paracompleteness (points next to **A** or **A** itself), or both big, characterizing inconsistency (points next to **B** or **B** itself). The equation of line **AB** is *a − b* = 0.

As we move away from the PUL, the indefinition gradually decreases, that is, certainty gradually increases. Moving away from the PUL, in the direction of point **C**, the favorable evidence increases and contrary evidence decreases, tending to a well defined situation of maximum certainty, of truth (point **C**). On the contrary, if we move away in the direction from the PUL to point **D**, the favorable evidence decreases and the contrary evidence increases, also tending to a well defined situation of minimum certainty, of falsity (point **D**).

Degree of certainty of an annotation (**a**; **b**) is defined as being

$$H(a; b) = a - b,$$

which translates, intuitively, the distance from point **X** = (**a**; **b**) *to line* **AB** (PUL), multiplied by $\sqrt{2}$ and affected by the signal + or − (here, metric distance was also considered, as in analytical geometry).

Observe that $- 1 \leq H \leq 1$.

Example H (0.9; 0.1) = 0.9 − 0.1 = 0.8. Therefore, the degree of certainty of the annotation constant (0.9; 0.1) is 0.8; therefore, it is positive and high, pointing out to a high degree of truth. On the other hand, H (0.1; 0.9) = 0.1 − 0.9 = −0.8; in this case, the degree of certainty is high (in module) and negative, pointing out to a high degree of falsity.

Each value of the degree of certainty belonging to the open interval]−1, 1 [defines a segment parallel to PUL. If $0 < H < 1$, this segment is called **truth limit line** (PQ, in Fig. 2.4); if $−1 < H < 0$, **falsity limit line** (TU, in Fig. 2.4); if $H = 0$, it is the perfectly undefined line (PUL); if $H = −1$ or 1, we have point **D** or point **C**, which are the extreme states of falsity and of truth, respectively.

Observe that the situation of "maximum truth" occurs in point C, when the favorable evidence is maximum ($a = 1$) and the contrary evidence is minimum ($b = 0$); in this case, the degree of certainty is maximum ($H = 1$).

The situation of "maximum falsity", in turn, occurs in point D, when the contrary evidence is maximum ($b = 1$) and the favorable evidence is null ($a = 0$). In this situation, the degree of certainty is minimum ($H = −1$).

2.6   The Lattice τ and the Decision States

The PDL divides the SUCP into two regions. This division tells if the situation represented by point $X = (a; b)$ is of paracompleteness (region **ACD**) or inconsistency (region **BCD**) (see Fig. 2.3).

An analogous comment is worth for the divisions of the SUCP done by the PUL in two regions or by bith, PDL and PUL, in four regions (Fig. 2.3).

When you wish to make a division of the SUCP so that the regions translate more precisely the analyzed situation (represented by point $X = (a; b)$), it must be done so in a greater number of regions. This may be done with certain criteria, which will be shown and commented. To begin with, the example shown in Fig. 2.4.

Observe that, in this division, besides the line segments **CD** (PDL): $a + b − 1 = 0$ and **AB** (PUL): $a − b = 0$, two parallels are used to each one of them, which are:

$$\text{Line } \mathbf{MN} : a + b − 1 = −0.6 \Rightarrow G = −0.6;$$
$$\text{Line } \mathbf{RS} : a + b − 1 = +0.6 \Rightarrow G = +0.6;$$
$$\text{Line } \mathbf{TU} : a − b = −0.6 \Rightarrow H = −0.6;$$
$$\text{Line } \mathbf{PQ} : a − b = +0.6 \Rightarrow H = +0.6.$$

With this division, four extreme regions and one central region may be highlighted.

$$\mathbf{AMN} \text{ Region: } − 1 \leq G \leq − 0.6 \quad \text{(paracompleteness state)}$$
$$\mathbf{BRS} \text{ Region: } 0.6 \leq G \leq 1 \quad \text{(inconsistency state)}$$

In these regions, we have situations of high indefinition. Therefore, if one point $X = (a; b)$, which translates a generic situation in study, belongs to one of them, we do not have a situation of high definition. On the contrary, they are situations of high indefinition: high paracompleteness (**AMN** region) or high inconsistency (**BRS** region).

Lines **MN** and **RS** are called **paracompleteness limit line** and **inconsistency limit line**, respectively. By convention, they belong to the analyzed regions.

Hence, these limit lines may be defined as follows:

Paracompleteness limit line (**MN**): $G = -k_1$ or $a + b - 1 = -k_1$, with $0 < k_1 < 1$;

Inconsistency limit line (**RS**): $G = -k_1$ or $a + b - 1 = k_1$, with $0 < k_1 < 1$;

In these definitions, the same value k_1 was used for both, giving symmetry to these lines. However, that is not mandatory. Different values could be adopted for k_1, k'_1 and k''_1, for the definition of each one.

CPQ Region: $0.6 \leq H \leq 1$ (truth region)
DTU Region: $-1 \leq H \leq -0.6$ (falsity region)

In contrast to the previous ones, in these regions, we have situations of high definition (truth or falsity). Hence, if point $\mathbf{X} = (a; b)$ belongs to one of these regions, we have: degree of certainty next to 1, characterizing a truth (**CPQ**: truth region) or degree of certainty next to -1, characterizing a falsity (**DTU**: falsity region).

With the concepts presented above, we start working with "truth region" instead of the "truth" being an inflexible concept. In this work, intuitively, truth is a region of high certainty, favorable in relation to the considered proposition; falsity is a region of high certainty, contrary in relation to the considered proposition.

Observe, then, that a great certainty (high value of the degree of certainty H module) means truth, if H is next to 1, and falsity, if H is next to -1.

Segments **PQ** and **TU** are called, respectively, of **truth limit line** and **falsity limit line**, which, by convention, belong to the analyzed regions.

Hence, these limit lines may be defined as follows:

Truth limit line (**PQ**): $H = k_2$ or $a - b = k_2$, with $0 < k_2 < 1$;

Falsity limit line (**TU**): $H = -k_2$, or $a - b = -k_2$, with $0 < k_2 < 1$.

In these definitions, the same value k_2 was used for both, giving symmetry to these lines. However, that is not mandatory. Different values could be adopted for k_2, k'_2 and k''_2, for the definition of each one.

In this book, $k_1 = k_2 = k$ will be adopted, giving symmetry to the chart, as in Fig. 2.4, in which we have $k_1 = k_2 = k = 0.60$. The value of k_2 will be called **requirement level**, as it represents the minimum value of $|H|$ so that point $\mathbf{X} \equiv (a; b)$ belongs to the falsity or truth region.

When a more strict criterion is desired, that is, greater precision for the conceptualization of truth or of falsity, it is sufficient to approximate lines **PQ** and **TU** to **C** and **D**, respectively, increasing the value of k_2, that is the requirement level. Observe that the requirement level will be stipulated for each application, "calibrating" the decision device according to the peculiarities the application presents.

It must be observed, for example, that when point $\mathbf{X} = (a; b)$ is internal to region **CPQ** (see Fig. 2.4), the truth will be defined with high degree of certainty ($0.6 \leq$ H ≤ 1), but with a small degree of uncertainty ($-0.4 \leq$ G ≤ 0.4). Therefore, as already previously mentioned, it is a logic that enables analyses that take into account the inconsistencies in the information, and accepts them until an established limit.

MNTUSRQP Region: $-0.6 < G < 0.6$ and $-0.6 < H < 0.6$

This is a region of low definition, as in it, G and H are small. Below, as an example, a detailed analysis of one of its sub-regions.

OFSL sub-region: $0.5 \leq a < 0.8$ and $0.5 \leq b \leq 1; 0 \leq G < 0.6$ and $-0.5 \leq H < 0$

In this sub-region we have a situation of relatively small inconsistency and falsity, but closer to the situation of total inconsistency (point **B**) than to the situation of total falsity (point **D**). For that reason, this sub-region is defined as of **quasi-inconsistency tending to falsity**.

Observe that the SUCP divided into these twelve regions, enables us to analyze the logical state of a proposition of the logic Eτ represented by point $\mathbf{X} = (a; b)$. That is the reason why this configuration was considered to **para-analyzer algorithm** [37] (Fig. 2.5).

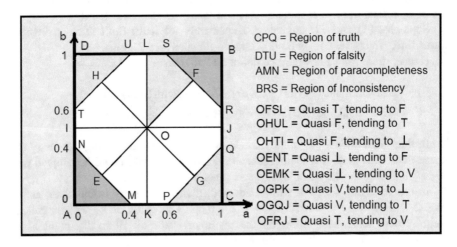

Fig. 2.5 Analysis of some regions of the SUCP

Table 2.1 Summary of Analysis of twelve regions of the unit square of the cartesian plane (SUCP)

Region	a	b	G	H	Description	Representation
AMN	[0; 0.4]	[0; 0.4]	[−1; −0.6]	[−0.4; 0.4]	Paracompleteness	⊥
BRS	[0.6; 1]	[0.6; 1]	[0.6; 1]	[−0.4; 0.4]	Inconsistency	⊤
CPQ	[0.6; 1]	[0; 0.4]	[−0.4; 0.4]	[0.6; 1]	Truth	V
DTU	[0; 0.4]	[0.6; 1]	[−0.4; 0.4]	[−1; −0.6]	Falsity	F
OFSL	[0.5; 0.8 [	[0.5; 1]	[0; 0.6 [	[−0.5; 0 [	Quasi inconsistency tending to falsity	QT → F
OHUL	] 0.2; 0.5 [	[0.5; 1]	[0; 0.5 [	]−0.6; 0 [	Quasi falsity tending to inconsistency	QF → T
OHTI	[0; 0.5 [	[0.5; 0.8 [	[−0.5; 0 [	] −0.6; 0 [	Quasi falsity tending to paracompleteness	QF → ⊥
OENI	[0; 0.5 [	] 0.2; 0.5 [	] −0.6; 0 [	] −0.5; 0 [	Quasi paracompleteness tending to falsity	Q⊥ → F
OEMK	] 0.2; 0.5 [	[0; 0.5 [	] −0.6; 0 [	[0; 0.5 [	Quasi paracompleteness tending to truth	Q⊥ → V
OGPK	[0.5; 0.8 [	[0; 0.5 [	[−0.5; 0 [	[0; 0.6 [	Quasi truth tending to paracompleteness	QV → ⊥
OGQJ	[0.5; 1]	] 0.2; 0.5 [	[0; 0.5 [	[0; 0.6 [	Quasi truth tending to inconsistency	QV → ⊤
OFRJ	[0.5; 1]	[0.5; 0.8 [	[0; 0.6 [	[0; 0.5]	Quasi inconsistency tending to truth	QT → V

Table 2.1 presents a summary of the logical state in each one of the four extreme regions and in the eight sub-regions highlighted in this analysis, which represent the logical decision states.

The representation of the lattice $\tau = \; <|\tau|, \; \leq^* >$ in the cartesian plane has already been seen with detail, placing on the abscissas axis the values of the degree of favorable evidence (a) and on the ordinates axis the values of the degree of contrary evidence (b). The so-called unit square of the cartesian plane (SUCP) and para-analyzing algorithm were obtained.

However, another representation of the referred lattice may be obtained placing on the abscissas axis the values of the degree of certainty (H) and on the ordinates axis the values of the degree of uncertainty (G). In this case, a square (which is not unitary) is also obtained, as shown in Fig. 2.6.

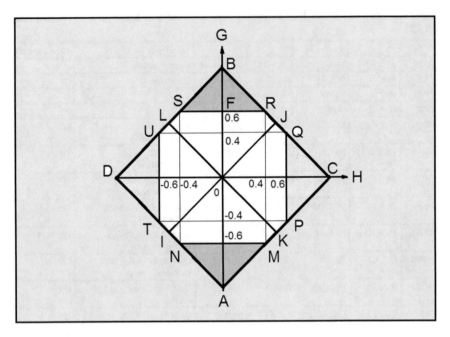

Fig. 2.6 Representation of the lattice associated to Eτ with the degrees of certainty on x-axis and the degrees of contradiction on y-axis

2.7 Logical Eτ Operators (NOT, MAX and MIN)

In the following paragraphs, the **NOT**, **MAX** and **MIN** operators on the lattice $\tau = \; <[0, 1] \times [0, 1], \leq^* >$ will be defined.

The operator **NOT** is defined by **NOT** $(a; b) = (b; a)$. We have used another symbol for this operator above.

In fact: $\sim (a; b) = (b; a)$ and **NOT**$(a; b) = (b; a)$.

Numerically: $\sim (0.8; 0.3) = (0.3; 0.8)$ or **NOT**$(0.8; 0.3) = (0.3; 0.8)$.

Observe that: **NOT**$(\top) = \top$; **NOT**$(\bot) = \bot$; **NOT**$(V) = F$ and **NOT**$(F) = V$.

An example: Suppose that the proposition **p** is "The Brazilian Soccer Team will qualify for the 2014 World Cup", with favorable evidence (degree of belief) equal to 0.8 (**a**) and contrary evidence (degree of disbelief) equal to 0.3 (**b**). Thus, we have: $\mathbf{p}_{(0.8; \; 0.3)}$.

Intuitively, its negation is the proposition: "The Brazilian Soccer Team will qualify for the 2010 World Cup", with favorable evidence equal to 0.3 (**b**) and contrary evidence equal to 0.8 (**a**).

Then, $\neg \mathbf{p}_{(0.8; \, 0.3)} = \mathbf{P}_{(0.3; \, 0.8)} = \mathbf{P}_{[\sim (0.8; \, 0.3)]}$.

The **MAX** operator (that will be called **maximizing**) of lattice $< [0. \, 1 \,] \times [0. \, 1]$, $\leq^* >$, associated to the logic Eτ, must be applied to a group of **n** annotations

$(n \geq 1)$. It acts in order to **maximize** the degree of certainty $(H = a - b)$ of this group of annotations, selecting the best favorable evidence (highest value of a) and the best contrary evidence (lowest value of b). It is defined as follows:

$$(a_1; b_1)\mathbf{MAX}(a_2; b_2)\mathbf{MAX}\ldots\mathbf{MAX}(a_n; b_n)$$
$$= (\max\{a_1, a_2, \ldots, a_n\}; \min\{b_1, b_2, \ldots, b_n\})$$

or, utilizing another representation:

$$\mathbf{MAX}\{(a_1; b_1), (a_2; b_2), \ldots (a_n; b_n)\} = (\max\{a_1, a_2, \ldots, a_n\}; \min\{b_1, b_2, \ldots, b_n\})$$

Therefore, to obtain the annotation resultant from the application of the **MAX** operator to the two annotations $(a_1; b_1)$ and $(a_2; b_2)$, the maximization between the values of the favorable evidence is made and, then, the minimization between the values of the contrary evidence. The resultant values of the evidences, favorable and contrary, are, respectively:

$$a' = \max\{a_1, a_2\} \quad \text{and} \quad b' = \min\{b_1, b_2\}$$

Observe that the **MAX** operator acts in order to choose, within the set of annotations, a degree of favorable evidence and a degree of contrary evidence, in such a manner that point $X = (a'; b')$ of the SUCP defined by the resultant pair $(a'; b')$ is the closest to point $\mathbf{C}$, extreme (or cardinal) point of truth (see Fig. 2.7).

The **MIN** operator (that will be called **minimizing**) of lattice $< [0, 1] \times [0, 1], \leq^* >$, associated to the logic Eτ, must be applied to a group of **n** annotations $(n \geq 1)$. It acts in order to **minimize** the degree of certainty $(H = a - b)$ of this group of annotations, selecting the worst favorable evidence (lowest value of a) and the worst contrary evidence (highest value of b). It is defined as follows:

$$(a_1; b_1)\mathbf{MIN}(a_2; b_2)\mathbf{MIN}\ldots\mathbf{MIN}(a_n; b_n) = (\min\{a_1, a_2, \ldots, a_n\}; \max\{b_1, b_2, \ldots, b_n\})$$

or, utilizing another representation:

$$\mathbf{MIN}\{(a_1; b_1), (a_2; b_2), \ldots (a_n; b_n)\} = (\min\{a_1, a_2, \ldots, a_n\}; \max\{b_1, b_2, \ldots, b_n\})$$

Therefore, to obtain the annotation resultant from the application of the **MIN** operator to the two annotations $(a_1; b_1)$ and $(a_2; b_2)$, the minimization between the values of the favorable evidence is made and, then, the maximization between the values of the contrary evidence. The resultant values of the evidences, favorable and contrary, are, respectively:

$$a'' = \min\{a_1, a_2\} \quad \text{and} \quad b'' = \max\{b_1, b_2\}$$

Observe that the **MIN** operator acts in order to choose, within the set of annotations, a degree of favorable evidence and a degree of contrary evidence, in

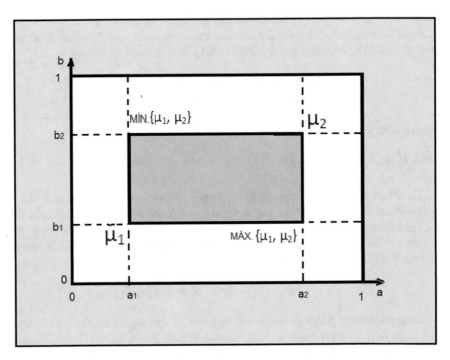

Fig. 2.7 Graphic interpretation of this application of the **MAX** and **MIN** operators

such a manner that point X = $(a''; b'')$ of the SUCP defined by the resultant pair $(a'';$ $b'')$ is the closest to point **D**, extreme (or cardinal) point of falsity (see Fig. 2.7).

Being $\mu_1 = (a_1; b_1)$, $\mu_2 = (a_2; b_2)$, $a_1 \leq a_2$ and $b_1 \leq b_2$, result that

$$\mathbf{MAX}\{\mu_1, \mu_2) = \mathbf{MAX}\{(a_1; b_1), (a_2; b_2)\} = (a_2; b_1) \text{ and}$$
$$\mathbf{MIN}\{\mu_1, \mu_2) = \mathbf{MIN}\{(a_1; b_1), (a_2; b_2)\} = (a_1; b_2).$$

The graphic interpretation of this application of the **MAX** and **MIN** operators is in Fig. 2.7.

In the applications of the **MAX** and **MIN** operators in real case studies, for the aid in the decision making, some details must be observed.

As it has already been seen, the **MAX** operator has the sense of performing the **maximization** of the degree of certainty for a set of annotations; therefore, it must be applied in situations in which the two or more considered items **are not all determining factors**, being sufficient that one of them has a favorable condition for the analysis result to be considered satisfactory.

The **MIN** operator has the sense of performing the **minimization** of the degree of certainty for a set of annotations; therefore, it must be applied in situations in which the two or more considered items **are all determining factors,** being essential that all of them present favorable conditions for the analysis result to be considered satisfactory.

Normally, what is done when designing an analysis of a real situation is to separate the researched items (or the experts) into groups [44]. They must be constituted in such a manner that:

(a) the existence of an item (or an expert) inside each group with favorable condition is sufficient to consider the result of the group as satisfactory;
(b) there are as many groups as the minimum number of items (or experts), which must have favorable conditions to consider the research result as satisfactory.

After this division is made, the **MAX** operator is applied inside each group (intragroup) and, then, the **MIN** operator between the results obtained in the groups (intergroup).

Below is the analysis of a simple example for a better comprehension. Imagine that a engineering structural construction work has presented some cracking. To verify the severity of the problem, the person responsible for the construction work collected the opinion of four engineers, specialists in soil mechanics (E_1), structures (E_2), civil construction materials (E_3) and foundations (E_4).

To analyze the engineers' opinions in view of the Eτ, a reasonable way to group them is to constitute a group A formed by the soil mechanics and foundations experts (E_1 and E_4) and another group B, composed by the specialists in structures and civil construction materials (E_2 and E_3).

Here we assume that, if the opinion of E_1, or E_4 is in the sense of stating that the problem is serious, that is sufficient to consider Group A favorable to the severity of the problem; analogous thought applies to Group B. However, to conclude by means of this analysis that the problem is really serious, it is required that both groups, A and B, give their opinion for the severity of the problem.

Thus, the maximization rule (**MAX** operator) is applied inside each group (intragroup) and the minimization rule (**MIN** operator) for the results obtained in the two groups (intergroup). The application of the rules, in this case, is this way:

$$[(E_1)\textbf{MAX}(E_4)]\textbf{MIN}[(E_2)\textbf{MAX}(E_3)] \quad \text{or} \quad [G_A]\textbf{MIN}[G_B]$$

or, utilizing another representation:

$$\textbf{MIN}\{\textbf{MAX}[(E_1),(E_4)]; \textbf{MAX}[(E_2),(E_3)]\} \quad \text{or} \quad \textbf{MIN}\{[G_A];[G_B]\}$$

This way of applying the maximization and minimization rules for decision making is known as **principle of the minimax** or optimistic decision, as it minimizes the highest degree of certainty.

Figure 2.8 represents the application of these rules in a schematic manner.

The application of these operators enables to determine possible inconsistencies of the database and verify to what extent they are acceptable or not in decision making.

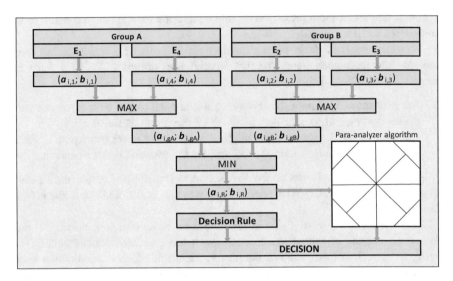

Fig. 2.8 Scheme for application of the **MAX** and **MIN** operators

The paraconsistent annotated evidential logic Eτ, although very recently discovered, has been finding applications in several activity fields. It is believed that the adequacy of the logic Eτ to these applications is due to the fact of enabling the work with knowledge bases containing inaccurate, inconsistent and paracomplete data, but not trivial data. In fact, most of the times, when a research is conducted among clients or suppliers or even among experts, the obtained information is vague or not always consistent, and we may even come across incomplete data. This way, to address a database with these characteristics, it is convenient to have a tool that is of simple application, efficient and, preferably, is easily computerized. And this is exactly the profile of the logic Eτ. Through it, we manage to analyze the data, although it is inaccurate, inconsistent or paracomplete, filter it and reach a final result that, analyzed in the lattice τ, will enable a conclusion.

Exercises

2.1 Given the propositions: $\mathbf{p} \equiv$ "Peter has a fever" and $\mathbf{q} \equiv$ "Peter must rest", express, utilizing convenient annotations in the language of the logic Eτ:

  (a) "Peter does not have a fever";
  (b) "Peter has a fever and does not have a fever";
  (c) "Peter must rest".

2.2 Given the annotation constants, calculate the degrees of certainty and contradiction. Verify which cardinal (or extreme) state each one of them is closest to,

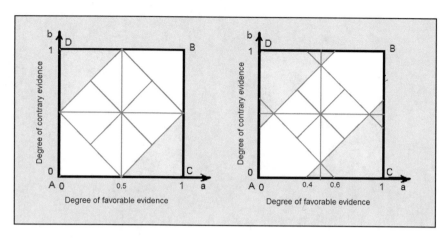

Fig. 2.9 Corresponding lattices

(a) (0.9; 0.2); (b) (0.1; 0.7); (c) (0.1; 0.2); (d) (0.8; 0.9); (e) (0.5; 0.5).

2.3 Make the figures of the para-analyzing algorithm (analogous to Fig. 2.4) with:

(a) requirement level 0.5; (b) requirement level 0.4.

2.4 Calculate:

(a) **MAX** [(0.9; 0.8), (0.5; 0.3), (0.1; 0.7)];
(b) **MIN** [(0.9; 0.8), (0.5; 0.3), (0.1; 0.7)];
(c) **MAX** {**MAX** [(0.9; 0.8), (0.5; 0.3), (0.1; 0.7)]; **MIN** [(0.9; 0.8), (0.5; 0.3), (0.1; 0.7)] };
(d) **MIN** {**MAX** [(0.9; 0.8), (0.5; 0.3), (0.1; 0.7)]; **MIN** [(0.9; 0.8), (0.5; 0.3), (0.1; 0.7)] }.

Answers

2.1 (a) $\mathbf{p}_{(0.0;\ 0.1)}$; (b) $\mathbf{p}_{(1;\ 1)}$; (c) $\mathbf{q}_{(1;\ 0)}$.
2.2 (a) G(0.9; 0.2) = 0.7; H(0.9; 0.2) = 0.1; Truth.
   (b) G(0.1; 0.7) = –0.6; H(0.1; 0.7) = –0.2; Falsity.
   (c) G(0.1; 0.2) = –0.1; H(0.1; 0.2) = –0.2; Paracompleteness.
   (d) G(0.8; 0.9) = –0.1; H(0.8; 0.9) = 0.7; Inconsistency.
   (e) G(0.5; 0.5) = 0; H(0.5; 0.5) = 0;
   Such annotation constant is equidistant from all the cardinal states.
2.3 (Fig. 2.9).
2.4 (a) (0.9; 0.3); (b) (0.1; 0.8); (c) (0.9; 0.3); (d) (0.1; 0.8).

Chapter 3
Decision Rules

3.1 General Considerations

A convenient division of lattice τ is seen in Fig. 3.1, in which the unit square is divided into twelve regions. Out of these, the four regions denominated extreme regions are highlighted, which will be object of more detailed analysis.

In this division of τ, we highlight the segments **AB**, called perfectly undefined line (PUL), and **CD**, called perfectly defined line (PDL). For a given annotation constant $(a; b)$, we will define the

- **degree of uncertainty**, by the expression $G(a; b) = a + b - 1$ (proportional to the distance from the point that represents it to the PDL); and also, the
- **degree of certainty**, by the expression $H(a; b) = a - b$ (proportional to the distance from the point that represents it to the PUL).

For points $X = (a; b)$ next to **A**, the values of the degree of favorable evidence (or degree of belief) (a) and of the degree of contrary evidence (or degree of disbelief) (b) are close to 0, characterizing a region of paracompleteness (**AMN**); next to **B**; on the contrary, the values of a and b are close to 1, characterizing a region of inconsistency (**BRS**); in the surroundings of **C**, the values of a are close to 1 and the values of b are close to 0, defining a region with high degree if certainty (close to 1), called truth region (**CPQ**); and, finally, in the proximities of **D**, the values of a are close to 0, and the values of b are close to 1, defining a region with low degree of certainty (close to −1), but high in module, called falsity region (**DTU**).

Also defined, are:

Paracompleteness limit line: line **MN**, so that $G = -k_1 = -0.70$, where $0 < k_1, < 1$;
Inconsistency limit line: line **RS**, so that $G = k_1 = 0.70$, where $0 < k_1 < 1$;
Falsity limit line: line analyses for decision making; for that reason, it, so that $H = -k_2 = -0.70$, where $0 < k_2 < 1$;
Truth limit line: line **PQ**, so that $H = k_2 = 0.70$, where $0 < k_2, < 1$.

© Springer International Publishing AG, part of Springer Nature 2018
F. R. de Carvalho and J. M. Abe, *A Paraconsistent Decision-Making Method*,
Smart Innovation, Systems and Technologies 87,
https://doi.org/10.1007/978-3-319-74110-9_3

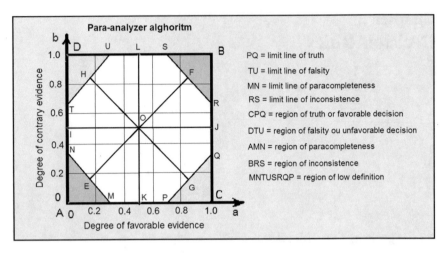

Fig. 3.1 Extreme regions with degrees of contradiction and of certainty, in module, equal or higher than 0.70

Except for contrary reference, in this book, $k_1 = k_2 = k$ will be adopted, giving symmetry to the chart, as in Fig. 3.1, in which $k_1 = k_2 = k = 0.70$. The value of k_2 will be called **requirement level** (control value), as it will be seen further ahead.

As seen in the previous chapter, four extreme regions and one central region are highlighted in Fig. 3.1.

AMN Region: $-1.0 \leq G \leq -0.70 \Rightarrow$ paracompleteness region

BRS Region: $0.70 \leq G \leq 1.0 \Rightarrow$ inconsistency region

In these regions, we have situations of high indefinition ('very' paracomplete or 'very' inconsistent). Therefore, if point $\mathbf{X} = (a; b)$, *which translates a generic situation in study, belongs to one of these regions, it will be said that the data present a high degree of uncertainty (paracompleteness or inconsistency).*

CPQ Region : $0.70 \leq H \leq 1.0 \Rightarrow$ truth region

DTU Region : $-1.0 \leq H \leq -0.70 \Rightarrow$ falsity region

In contrast to the previous ones, in these regions, we have situations of high definition (truth or falsity). Therefore, if point $\mathbf{X} = (a; b)$, which translates a generic situation in study, belongs to the region **CPQ** or **DTU**, it will be said that the situation presents a high degree of favorable certainty (truth) or contrary certainty (falsity), respectively.

The first one, **CPQ**, is called favorable decision region (or feasibility), as when the point that translates the result of the analysis belongs to it, it means that the result presents a high degree of favorable evidence (degree of belief) and low degree of contrary evidence (degree of disbelief). That results in a high degree of certainty (close to 1), which leads to a favorable decision, translating the feasibility of the enterprise.

The second one, **DTU**, is called unfavorable decision region (or unfeasibility), as, belonging to this region, the result presents a low degree of favorable evidence and high degree of contrary evidence. That results in a low degree of certainty (close to -1), which leads to an unfavorable decision, translating the unfeasibility of the enterprise.

MNTUSRQP Region

$$|G| < 0.70 \Rightarrow -0.70 < G < 0.70 \text{ and } |H| < 0.70 \Rightarrow -0.70 < H < 0.70$$

This is a region that does not allow highlighted conclusions, that is, when the point that translates the result of the analysis belongs to this region, it is not possible to say that the result has a high degree of certainty or uncertainty. This region translates only the tendency of the analyzed situation, according to the considered decision states (see Sect. 2.6 and Table 2.1).

This way, the favorable decision (feasibility) is made when the point that translates the analysis result belongs to the truth region (**CPQ**); and the unfavorable decision (unfeasibility), when the result belongs to the falsity region (**DTU**).

3.2 Requirement Level and the Decision Rule

As a result of the considerations made in the previous section, for the configuration in Fig. 3.1, we may enunciate the following decision rule [45]:

> $H \geq 0.70 =$ **favorable decision (the enterprise is feasible)**;
> $H \leq -0.70 =$ **unfavorable decision (the enterprise is unfeasible)**;
> $-0.70 < H < 0.70 =$ **inconclusive analysis**.

Observe that, for this configuration, the decision (favorable or unfavorable) is only made when $|H| \geq 0.70$, or that is, when $k_2 = 0.70$. Hence, this value ($k_2 = 0.70$) represents the lowest value of the degree of certainty module for which a decision is made. For that reason, it is here denominated **requirement level (NE)** of the decision. With that, the decision rule, represented in the most generic form, is this way:

H $\geq$ NE = favorable decision (the enterprise is feasible);

H $\leq$ $-$NE = unfavorable decision (the enterprise is unfeasible);

$-$ NE $<$ H $<$ NE = inconclusive analysis.

It is appropriate to highlight that the requirement level depends on the safety, on the trust that you wish to have in the decision, which, in turn, depends on the responsibility it implicates, on the investment at stake, on the involvement or not of risk to human lives, etc.

If we want a stricter criterion for the decision making, that is, if we want safer, more reliable decisions, it is required to increase the **requirement level**, that is, we must approximate lines **PQ** and **TU** of points **C** and **D**, respectively.

Observe that, if the result belongs to the **BRS** region (inconsistency region), the analysis is inconclusive regarding the feasibility of the enterprise, but it points out to a high degree of inconsistency of the data (G $\geq$ 0.70).

Analogously, if the result belongs to the **AMN** region (paracompleteness), it means that the data present a high degree of paracompleteness or, equivalently, high lack of information about the data (G $\leq$ –0.70).

In these cases, therefore, the result does not enable a conclusion regarding the feasibility of the enterprise, but it allows us to conclude that the database, which will often be constituted by experts' opinions, presents a high degree of uncertainty (paracompleteness or inconsistency). Therefore, it allowsus, at least, to have information regarding the degree of uncertainty of the elements contained in the database. This is a great advantage of using the paraconsistent annotated evidential logic Eτ, which manages to handle data, even if they are provided with paracompletenesses or inconsistencies (or contradictions).

As we may see, the application of the paraconsistent logic techniques enables us to determine possible inconsistencies of the database and verify to what extent they are acceptable or not in decision making.

The importance of the analysis of a real situation through **MAX** and **MIN** operators lies in the fact that it, even if the analyzed conditions present contradictory results, they are taken into account. That means that this method accepts databases that present contradictions, or paracompletenesses, that is, it manages to deal with situations provided with uncertainties, as long as they are not trivial. This is the great merit of the paraconsistent annotated logics.

It is verified, then, that lattice τ with the division into twelve logical decision states (Fig. 3.1) enables analyses for decision making; for that reason, it was considered to **para-analyzer algorithm (or device)** [37, 41].

Chapter 4
The Decision Making Process

Paraconsistent Decision Method (PDM)

4.1 Initial Considerations

Every reasonable decision must be based on a broad series of factors that may influence the enterprise in the analysis. Out of these factors, each one will influence differently, indicating by the feasibility (favorable decision) or by the unfeasibility (unfavorable decision) of the enterprise, or even, the factor may be inconclusive, not providing an indication, neither favorable not contrary. This is very noticeable when the para-analyzing algorithm is utilized, that is, when the values of the favorable evidence (or degrees of belief) $(a_{i,R})$ and contrary evidence (or degrees of disbelief) $(b_{i,R})$, resulting from each factor, are plotted, so that each factor is represented by one point $X = (a; b)$ of lattice τ.

However, working with a substantial number of factors is not reasonable, as it would make the method exhausting and costly. Thus, the proposal is choosing and utilizing only the most important ones, the ones with most influence in the decision, within the boundary of the limited rationality recommended by Simon, "which works with a simplified model of reality, considering that several aspects of reality are substantially irrelevant at a certain moment; the choice is made based on the satisfactory standard of the real situation, considering only some of the most relevant and crucial factors" [94].

Normally, knowing how the isolated influence of each factor does not present relevant interest. What matters in the feasibility analysis of an enterprise is the joint influence of all the chosen factors, which is translated by a final logical state denominated center of gravity (w). It is represented by a point W of lattice τ, whose coordinates $(a_w$ and $b_w)$ are determined by the weighted average of the coordinates of points $X_i = (a_{i,R}, b_{i,R})$ of τ, which translate the influence resulting from each one of the factors.

© Springer International Publishing AG, part of Springer Nature 2018
F. R. de Carvalho and J. M. Abe, *A Paraconsistent Decision-Making Method*,
Smart Innovation, Systems and Technologies 87,
https://doi.org/10.1007/978-3-319-74110-9_4

Thus:

$$a_w = \frac{\sum P_i \times a_{i,R}}{\sum P_i} \quad \text{and} \quad b_w = \frac{\sum P_i \times b_{i,R}}{\sum P_i} \tag{4.1}$$

where

P_i are the weights of the factors in the enterprise feasibility analysis.

Observe that the center of gravity W corresponds to the center of gravity of the points X_i that represent the factors isolatedly in lattice τ. If all the factors had the same weight in the decision, point W would coincide with the geometrical center of these points. Therefore, to obtain the final result of the analysis and make the decision, it is necessary to analyze the center of gravity of the points that represent the factors in lattice τ.

The Paraconsistent Decision Method (PDM) consists, basically, of eight stages listed here, and which will be detailed as the text follows.

1. Establish the requirement level of the decision you intend to make.
2. Select the most important factors and with the most influence in the decision.
3. Establish the sections for each one of the factors (three, four, five or more sections may be established, depending on the case and on the desired detailing).
4. Build the database, which is constituted by the weights attributed to the factors (when you wish to distinguish them by importance) and by the values of favorable evidence (or degree of belief) (*a*) and of contrary evidence (or degree of disbelief) (*b*) attributed to each one of the factors in each one of the sections; the weights and the values of the evidences are attributed by experts conveniently chosen to give their opinions. (The database may also be constructed with stored statistical data, obtained in previous experiences in the execution of similar enterprises).
5. Perform the field research (or data survey) to verify, in case of analysis, in which section (condition) each one of the factors is.
6. Obtain the value of the favorable evidence $(a_{i,R})$ and the value of the contrary evidence $(b_{i,R})$, resultant, with $1 \leq i \leq \mathbf{n}$, for each one of the chosen factors (F_i), in the sections found in the research (S_{pj}), by means of applications of the maximization (**MAX** operator) and minimization (**MIN** operator) techniques of the logic Eτ.
7. Obtain the degree of favorable evidence (a_w) and the degree of contrary evidence (b_w) of the center of gravity of the points that represent the chosen factors in lattice τ.
8. Make the decision, applying the decision rule or the para-analyzer algorithm.

4.2 Stages of the Paraconsistent Decision Method (PDM) [46]

To perform a feasibility analysis of an enterprise for decision making, the planning is under the coordination of a particular person (the businessperson him/herself, an engineer, a consultant, etc.). This person will work the data in such a manner as to "translate" them to the language of the logic Eτ, allowing, this way, a proper "plotting" to the analyses of the tools offered by this logic. This specialist will be called knowledge engineer (KE). Therefore, the knowledge engineer is responsible for organizing the whole analysis process for the final decision making.

4.2.1 Establishment of the Requirement Level

The first task of the KE is to establish the requirement level of the decision to be made. It is not hard to understand that the requirement level depends on the safety it is intended for the decision, which, in turn, depends on the responsibility it implicates (on the amount of the investment at stake, on the involvement or not of risk to human lives, etc.).

When the KE establishes the requirement level of the decision, he/she is automatically fixing the decision regions and, consequently, also the decision rule and the para-analyzer algorithm. In fact, for example, if the requirement level is fixed equal to 0.70 (that is, if it is established that the decision will only be made when the module of the difference between the degrees of favorable evidence and contrary evidence of the center of gravity is, at least, 0.70), the decision rule and the para-analyzer algorithm will be the ones represented in Fig. 4.1.

4.2.2 Choice of the Influence Factors

The second task of the KE is to research and verify the factors that influence the success (or failure) of the enterprise. This research encompasses all kinds of consultations he/she may conduct: consultation to personnel that act in an institution of the same field or similar field, to specialized compendia, to experts in the subject, consultation to other projects of the same enterprise or similar enterprises, etc.

After the factors that influence the success (or failure) of the enterprise have been researched, it is also the KE's responsibility to select the most important and most influential n factors F_i ($1 \leq i \leq n$), that is, the ones whose conditions may markedly affect the feasibility of the enterprise. One of these factors in very favorable conditions makes the analysis result tend highly to feasibility (favorable decision) of the enterprise. If the chosen factors may influence in different ways or

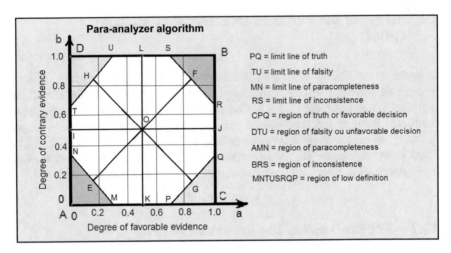

Fig. 4.1 Decision rule and para-analyzer algorithm, for requirement level equal to 0.70

if they have different importance in the decision, one more than the others, these differences may be compensated by the attribution of different weights to each one of the chosen factors.

4.2.3 Establishment of the Sections for Each Factor

The following mission of the KE is to establish the **sections** S_j $(1 \leq j \leq s)$, which translate the conditions in which each factor may be found. Then, depending on the refinement he/she intends to provide the analysis, more (or less) sections may be established.

If the KE opts for establishing three sections, they will be:

S_1—the factor is in **favorable** condition to the enterprise;
S_2—the factor is in **indifferent** condition to the enterprise;
S_3—the factor is in **unfavorable** condition to the enterprise.

For example, when analyzing the feasibility of a higher education institution to open a new course in a certain region, and the factor is the **amount of the course monthly fee** (M_c), the three established sections may be:

S_1: $M_c < 80\%\ M_m$; S_2: $80\%\ M_m \leq M_c \leq 120\%\ M_m$; S_3: $M_c > 120\%\ M_m$, being M_m the average amount of the monthly fees of the same course (or similar courses) in the other schools of the same region.

If the knowledge engineer opts for establishing five sections, they will be:

S_1—the factor is in **very favorable** condition to the enterprise;
S_2—the factor is in **favorable** condition to the enterprise;
S_3—the factor is in **indifferent** condition to the enterprise;
S_4—the factor is in **unfavorable** condition to the enterprise;
S_5—the factor is in **very unfavorable** condition to the enterprise.

For example, when analyzing the feasibility of the launch of a new product and the factor is the **product price (Q) in the market**, the five sections to be established may be:

S_1: $Q < 70\%$ P;
S_2: 70% P $\leq Q < 90\%$P;
S_3: 90%P $\leq Q \leq 110\%$P;
S_4: 110%P $< Q \leq 130\%$P;
S_5: $Q > 130\%$P,

being P the average price of the same product (or similar products) already existent in the market.

4.2.4 Construction of the Database

An important task to be executed by the KE is the construction of the database. For this purpose, he/she must choose m experts E_k ($1 \leq k \leq m$), from the area or related areas. Each one of the chosen experts will utilize their knowledge, experiences, sensitivity, intuition, common sense, etc., to provide information about the possibilities of the enterprise in the conditions established by the sections, for each one of the chosen influence factors.

In the choice of the experts, if possible, we must search for people with different majors, so that the attribution of values does not result from one single line of thought.

For example, a group of specialists adequate to analyze the feasibility of a higher education institution to open a new course in a certain region is composed by a professional with a sociologist formation, an economist, an educator, and a last one with a major in business administration (or school administration).

Evidently, it is possible to increase the comprehensiveness of the process about the experts' majors. For this purpose, just utilize more than four specialists and/or professionals from other major areas. There is nothing to prevent you from using more than one expert from the same major area. The KE is in charge of deciding about the choice of the experts.

It must be pointed out that the process presents great versatility, once it allows the choice of more (or less) influence factors, allows the establishment of three or more sections for each factor, as well as the utilization of a larger (or smaller) number of specialists. Although the process allows, it is not advisable to utilize less than four specialists, so that the result is not provided with too much subjectivity.

Firstly, the experts will say if, among the chosen factors, there is distinction regarding the importance. If there is not, they will be attributed the same weight (which may be equal to 1 (one) for all); if there is, each expert will attribute to each factor the weight $q_{i,k}$ they deem adequate, taking into account the importance of the factor in relation to the others in the decision that will be made.

$q_{i,k}$ = weight attributed by specialist k to factor i.

In the attribution of these weights, some restrictions may be imposed by the KE. For example, it may be imposed that the weights must be positive integers and which belong to the interval [1, 10]. After the weights have been attributed to all the factors by all the invited specialists, the arithmetic average of the weights attributed by the specialists will be adopted as the final weight P_i of each factor.

$$P_i = \frac{\sum_{k=1}^{m} q_{i,k}}{m} = \text{weight of the i due to all the specialists} \qquad (4.2)$$

Observe that there is the possibility that the KE may distinguish the experts according to the background (practice, experience, knowledge) of each one, attributing different weights r_k to the experts. In this case, the final weight P_i of each factor would not be an arithmetic average anymore, and would be a weighted average.

r_k = weight attributed by the engineering knowledge to specialist k.

$$P_i = \frac{\sum_{k=1}^{m} r_k \cdot q_{i,k}}{\sum_{k=1}^{m} r_k} = \text{weight of the i due to all the specialists} \qquad (4.3)$$

This is only one nuance of the method, showing its versatility and great quantity of options given to the Knowledge Engineer (KE).

After the final weights of the factors have been established, the specialists must be requested to attribute the value of the favorable evidence (or degree of belief) (**a**) and contrary evidence (or degree of disbelief) (**b**) to each one of the factors in the conditions in which it may be, characterized by the established sections. Of course, in these attributions, each specialist must also make use of their knowledge, experiences, sensitivity, intuition, common sense, etc.

Each ordered pair $(a_{i,j,k};\ b_{i,j,k})$ formed by the values of favorable and contrary evidences, attributed by a specialist E_k to a factor F_i inside the condition defined by a section S_j, constitutes an annotation constant symbolized by $\mu_{i,j,k}$.

The database is built by the matrix of the weights, M_p, column matrix $[P_i]$ of n lines formed by the average weights P_i of the factors; and by the matrix of the annotations, M_A, the matrix $[\mu_{i,j,k}]$ of n × s lines and m columns, that is, with n × s × m elements (bi-valued annotations), formed by all the annotations the m specialists attribute to each one of the n factors inside the conditions defined by the s sections.

The matrix $[\mu_{i,j,k}]$ may be represented by $[(a_{i,j,k};\ b_{i,j,k})]$, once each one of its elements $\mu_{i,j,k}$ is one ordered pair of the form $(a_{i,j,k};\ b_{i,j,k})$.

Table 4.1 Database: matrices of the weights, M_p, and of the annotations, M_A

			E_1	E_2	E_3	E_4
Fator F_1	P_1	Seção S_1	$\mu_{1.1.1}$	$\mu_{1.1.2}$	$\mu_{1.1.3}$	$\mu_{1.1.4}$
		Seção S_2	$\mu_{1.2.1}$	$\mu_{1.2.2}$	$\mu_{1.2.3}$	$\mu_{1.2.4}$
		Seção S_3	$\mu_{1.3.1}$	$\mu_{1.3.2}$	$\mu_{1.3.3}$	$\mu_{1.3.4}$
Fator F_2	P_2	Seção S_1	$\mu_{2.1.1}$	$\mu_{2.1.2}$	$\mu_{2.1.3}$	$\mu_{2.1.4}$
		Seção S_2	$\mu_{2.2.1}$	$\mu_{2.2.2}$	$\mu_{2.2.3}$	$\mu_{2.2.4}$
		Seção S_3	$\mu_{2.3.1}$	$\mu_{2.3.2}$	$\mu_{2.3.3}$	$\mu_{2.3.4}$
Fator F_3	P_3	Seção S_1	$\mu_{3.1.1}$	$\mu_{3.1.2}$	$\mu_{3.1.3}$	$\mu_{3.1.4}$
		Seção S_2	$\mu_{3.2.1}$	$\mu_{3.2.2}$	$\mu_{3.2.3}$	$\mu_{3.2.4}$
		Seção S_3	$\mu_{3.3.1}$	$\mu_{3.3.2}$	$\mu_{3.3.3}$	$\mu_{3.3.4}$
Fator F_4	P_4	Seção S_1	$\mu_{4.1.1}$	$\mu_{4.1.2}$	$\mu_{4.1.3}$	$\mu_{4.1.4}$
		Seção S_2	$\mu_{4.2.1}$	$\mu_{4.2.2}$	$\mu_{4.2.3}$	$\mu_{4.2.4}$
		Seção S_3	$\mu_{4.3.1}$	$\mu_{4.3.2}$	$\mu_{4.3.3}$	$\mu_{4.3.4}$
Fator F_5	P_5	Seção S_1	$\mu_{5.1.1}$	$\mu_{5.1.2}$	$\mu_{5.1.3}$	$\mu_{5.1.4}$
		Seção S_2	$\mu_{5.2.1}$	$\mu_{5.2.2}$	$\mu_{5.2.3}$	$\mu_{5.2.4}$
		Seção S_3	$\mu_{5.3.1}$	$\mu_{5.3.2}$	$\mu_{5.3.3}$	$\mu_{5.3.4}$

For example, in a situation with four specialists ($m = 4$), five factors ($n = 5$) and three sections for each factor ($s = 3$), the matrix of the weights, M_p, is a column matrix of 5 lines ($n = 5$) and the matrix of the annotations, M_A, has the form 15×4 ($n \times s = 5 \times 3 = 15$ e $m = 4$) as indicated in Table 4.1.

4.2.5   Field Research

The measures were taken by the KE until the construction of the database (from 1 to 4) complete his/her decision-making device. From that on, he/she is apt to apply the method and reach to the final decision, using information that will be collected using research about the condition (defined by the section) of each influence factor. Therefore, the following step is to conduct the field research and verify the real condition of each one of the influence factors, that is, to research in which Section S_j each Factor F_i is.

Thus, using the previous example (Sect. 4.2.3) of the opening of a course by an education institution, if the monthly fee M_c needed for the healthy functioning of the course is equal to $130\% M_m$, we say that the factor "**amount of the monthly course fee**" is in unfavorable condition for the opening of the course, that is, that the condition of this factor is translated by Section S_3 or that this factor is in section S_3.

Analogously, if in the other example seen in the same Sect. 4.2.3, the research shows that the price Q of launch of the product is $80\% P$, we say that the factor

Table 4.2 Matrices of the weights, M_{pi}, researched, M_{pq}, and of the researched data, M_{Dpq}

			E_1	E_2	E_3	E_4
Fator F_1	P_1	S_{p1}	$\lambda_{1.1}$	$\lambda_{1.2}$	$\lambda_{1.3}$	$\lambda_{1.4}$
Fator F_2	P_2	S_{p2}	$\lambda_{2.1}$	$\lambda_{2.2}$	$\lambda_{2.3}$	$\lambda_{2.4}$
Fator F_3	P_3	S_{p3}	$\lambda_{3.1}$	$\lambda_{3.2}$	$\lambda_{3.3}$	$\lambda_{3.4}$
Fator F_4	P_4	S_{p4}	$\lambda_{4.1}$	$\lambda_{4.2}$	$\lambda_{4.3}$	$\lambda_{4.4}$
Fator F_5	P_5	S_{p5}	$\lambda_{5.1}$	$\lambda_{5.2}$	$\lambda_{5.3}$	$\lambda_{5.4}$

"product price in the market" is in favorable condition for the launch of the product, that is, that the condition of this factor is translated by Section S_2 or that this factor is in section S_2.

After this research is conducted, the KE obtains a set of n sections resultant from the research, $S_{p,i}$ with $1 \leq i \leq \mathbf{n}$, one for each factor, which translate in which real conditions the factors are. These n values of the sections resulting from the research constitute a column matrix of n lines, which will be called researched matrix, $M_{pq} = [S_{pi}]$. With this selection of the sections, performed through the field research, it is possible to search in the database the experts' opinions about the feasibility of the enterprise in the conditions in which the factors are. These opinions are translated by the values of the favorable and contrary evidences attributed to the factors in the conditions of the sections obtained in the research.

This way, from the matrix M_A of the database, another matrix may be highlighted, its subset, which will be called matrix of the researched data, $M_{Dpq} = [\lambda_{i,k}]$, of n lines and m columns, constituted by the lines of M_A corresponding to Sections S_{pi} obtained in the research.

4.2.6   Calculation of the Resulting Annotations

Besides the result obtained in the research (matrix of the researched data, M_{Dpq}), for the application of the logic $E\tau$ techniques, a measure must be taken by the KE: divide the specialists into groups according to criteria that must be defined by him/herself. That is, how will the specialist groups be constituted?

In the constitution of the groups of specialists for the application of the MAX and MIN operators in real case studies, with the aid in the decision making, some details must be observed.

As seen in Sect. 2.7, the MAX operator has the sense of performing the maximization of the degree of certainty inside a set of annotations, choosing the highest degree of favorable evidence and the lowest degree of contrary evidence. Therefore, it must be applied in situations when the opinions of two or more experts (or researched items) are not all determining factors, with just the favorable opinion of one of them being enough to consider the group's result satisfactory. Thus, if among the experts there is one that deserves special prominence in the subject, he/she must be alone in one group, so that his/her opinion is necessarily considered. On the other hand, if there are two specialists of the same level and acting in the same area or related areas, both of them may be placed in the same group, as if the opinion of one of them is satisfactory, it is already sufficient to consider this group's opinion as favorable to the enterprise.

The MIN operator has the sense of performing the minimization of the degree of certainty inside a set of annotations, choosing the lowest degree of favorable evidence and the highest degree of contrary evidence (see Sect. 2.7). Therefore, it must be applied in situations when the opinions of two or more experts (or researched items) are all determining factors, being essential that all of them be favorable so that the analysis result may be considered satisfactory.

Thus, for example, if among the specialists there are two high-level ones, they must be in different groups and alone, so that both their opinions are necessarily considered and not shared with others' opinions.

An example that may clarify the formation of the groups is the following. Imagine the four sectors of a soccer team: the goalkeeper (a player with number 1), the defense (four players numbered from 2 to 5), the midfield (three players numbered from 6 to 8) and the attack (three players numbered from 9 to 11). And what the soccer players call 4–3–3 scheme. A coach (here, the knowledge engineer) understands that, for the team to be great, it must have one great player in each sector, that is, a great goalkeeper, a great defender, a great midfielder and a great attacker.

Thus, in a feasibility analysis of the team, the groups are already naturally constituted. The goalkeeper, who is the only one in the sector, constitutes one group (A); the four defense players constitute another group (B), as just one of them being great is sufficient to meet the requirement of a great team; analogously, the three of the midfield constitute the third group (C) and the three attackers, the fourth group (D).

The distribution of the groups for the application of the MAX and MIN operators is the following:

$$\text{MIN}\{[Group\,A],\ [Group\,B],\ [Group\,C],\ [Group\,D]\}\ \text{or}$$
$$\text{MIN}\{[1],\ \text{MAX}\,[2,3,4,5],\ \text{MAX}\,[6,7,8],\ \text{MAX}\,[9,10,11]\}\ \text{or}$$
$$\text{MIN}\,[(a_A;b_A)],\ [(a_B;b_B)],\ [(a_C;b_C)],\ [(a_D;b_D)]\},$$

which may be represented by the scheme of Fig. 4.2.

Observe that the influence of the goalkeeper is very accentuated, as he determines the result of group A alone.

For further details regarding this example, see Chap. 9.

As another illustrative example, imagine that there is the suspicion that one individual has a certain disease, because he has been presenting some abdomen pain symptoms. To conduct an analysis employing the para-analyzer, the opinions of

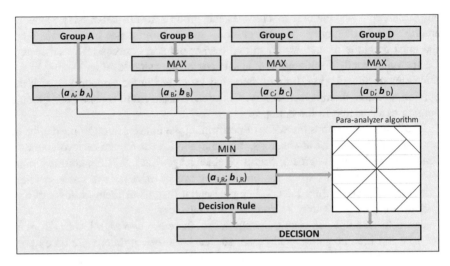

Fig. 4.2 Scheme of application of the **MAX** and **MIN** operators

four specialists (doctors) are collected: a primary care physician (E_1), a gastroenterologist (E_2), a urologist (E_3) and an endoscopist (E_4).

In a situation like this, suppose that the primary care physician's (E_1) and the endoscopist's (E_4) opinions are indispensable and crucial, but that the gastroenterologist's (E_2) and the urologist's (E_3) opinions do not have the same weight, and the opinion of just one of them may prevail.

In this situation, naturally, we already have a proper formation for the groups: group A, formed by the primary care physician (E_1); group B, by the endoscopist (E_4); and group C, formed by the gastroenterologist (E_2) and urologist (E_3). After we have the groups, the maximizing operator (**MAX**) is applied intragroup (which, in this case, is summarized to applying it to group C), and then, the minimizing operator (**MIN**) is applied among groups A, B and C (intergroups). In this case, the application of the operators is:

$$\mathbf{MIN}\{[E_1,], [E_4], \mathbf{MAX}[E_2, E_3]\} \quad \text{or} \quad \mathbf{MIN}\{[G_A],[G_B],[G_c]\}$$

In that case, a schematic representation of the application of the maximization and minimization rules is translated by Fig. 4.3.

It must be observed that the presented examples only have the purpose of illustrating the way the distribution of the researched items into groups and the application of the **MAX** and **MIN** operators is performed, and any discussion here regarding their technical correctness would go beyond the scope of this work.

The importance of the analysis of a real situation through **MAX** and **MIN** operators lies in the fact that it, even if the analyzed conditions present contradictory results, they are taken into account. That means that this method accepts

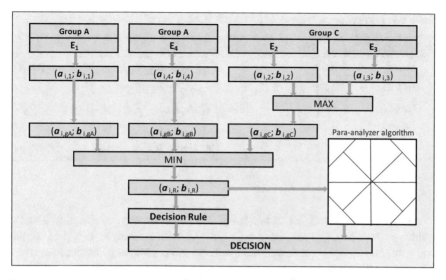

Fig. 4.3 Scheme for application of the **MAX** and **MIN** operators

databases that present contradictions, that is, it manages to deal with uncertain (inconsistent or paracomplete) situations.

The application of these operators enables us to determine the values of favorable evidence $(a_{i,R})$ and contrary evidence $(b_{i,R})$, with $1 \leq i \leq \mathbf{n}$, resultant, for each factor F_i in Section S_{pj} found in the research.

Recalling that every annotation (μ or λ) is, in this work, an ordered pair of the form $(a; b)$, we conclude that the matrix $M_{Dpq} = [\lambda_{i,k}]$ of the researched data may be represented by $[(a_{i,k}; b_{i,k})]$. This way, Table 4.2 starts having the form of Table 4.3.

Suppose that the knowledge engineer (KE) distributes the specialists into p G_h groups, with $1 \leq h \leq \mathbf{p}$, each one with g_h experts, being

$$\sum_{h=1}^{p} g_h = \mathbf{m}.$$

Thus, the G_h group will be constituted by the following g_h experts: E_{1h}, E_{2h}, ..., E_{ghh}. Then, the application of the maximization rule inside G_h group (intragroup) may be structured this way:

$$\mathbf{MAX}[(E_{1h}), (E_{2h}), \ldots, (Eg_{hh})] \text{ or}$$
$$\mathbf{MAX}[(a_{i,1h}; b_{i,1h}), (a_{i,2h}; b_{i,2h}), \ldots, (a_{i,ghh}; b_{i,ghh})]$$

As a result of this maximization we obtain the ordered pair $(\mathbf{a_{i,h}}; \mathbf{b_{i,h}})$, in which $a_{i,h} = \max. \{a_{i,1h}, a_{i,2h}, \ldots, a_{i,ghh}\}$ and $b_{i,h} = \max. \{b_{i,1h}, b_{i,2h}, \ldots, b_{i,ghh}\}$

Table 4.3 Matrices of the weights, M_{pi}, researched, M_{pq}, and of the researched data, M_{Dpq}

			E_1	E_2	E_3	E_4
Fator F_1	P_1	S_{p1}	$(a_{1,1}; b_{1,1})$	$(a_{1,2}; b_{1,2})$	$(a_{1,3}; b_{1,3})$	$(a_{1,4}; b_{1,4})$
Fator F_2	P_2	S_{p2}	$(a_{2,1}; b_{2,1})$	$(a_{2,2}; b_{2,2})$	$(a_{2,3}; b_{2,3})$	$(a_{2,4}; b_{2,4})$
Fator F_3	P_3	S_{p3}	$(a_{3,1}; b_{3,1})$	$(a_{3,2}; b_{3,2})$	$(a_{3,3}; b_{3,3})$	$(a_{3,4}; b_{3,4})$
Fator F_4	P_4	S_{p4}	$(a_{4,1}; b_{4,1})$	$(a_{4,2}; b_{4,2})$	$(a_{4,3}; b_{4,3})$	$(a_{4,4}; b_{4,4})$
Fator F_5	P_5	S_{p5}	$(a_{5,1}; b_{5,1})$	$(a_{5,2}; b_{5,2})$	$(a_{5,3}; b_{5,3})$	$(a_{5,4}; b_{5,4})$

As there are **n** factors, we obtain **n** ordered pairs this way, constituting a matrix resulting from the G_h group, $M_{Gh} = [(a_{i,h}; b_{i,h}]$, with **n** lines, as $1 \leq i \leq$ **n**, and one column. It must be observed that, as there are **p** groups, **p** matrices similar to this one are obtained.

Returning to the illustrative example of **n** = 5 factors, **s** = 3 sections and **m** = 4 specialists, and admitting that the four specialists were distributed into two groups (p = 2), the first one, G_p, by specialists E_1, and E_4, and the second one, G_2, by specialists E_2 and E_3, the application of the maximization rule would be performed the following way:

inside Group G_1: MAX $[(E_1,), (E_4)]$;
inside Group G_2: MAX $[(E_2), (E_3)]$ or
MAX $[(a_{i,1}; b_{i,1}), (a_{i,4}; b_{i,4})]$, resulting in $(a_{i,g1}; b_{i,g1})$ to group G_1; and
MAX $[(a_{i,2}; b_{i,2}), (a_{i,3}; b_{i,3})]$, resulting in $(a_{i,g2}; b_{i,g2})$ to group G_2, being

$$a_{i,g1} = \max \{a_{i1}, a_{i4}\}; \quad b_{i,g1} = \min \{b_{i1}, b_{i4}\}; \text{ and}$$
$$a_{i,g2} = \max \{a_{i2}, a_{i3}\}; \quad b_{i,g2} = \min \{b_{i2}, b_{i3}\};$$

We obtain, then, p = 2 column matrices with n = 5 lines as a result of the application of the maximization rule inside the groups G_1 and G_2 (intragroup). They are:

$$M_{G1} = [(a_{i,g1}; b_{i,g1})] = [\rho_{i,g1}] \text{ and } M_{G2} = [(a_{ig2}; b_{ig2})] = [\rho_{i,g2}],$$

which may be represented in another manner, as in Table 4.4.

After the maximization rules are applied (MAX operator) inside the groups (intragroup), the next step is the application of the minimization rule (MIN operator) among the p groups (intergroups), which may be structured this way:

Table 4.4 Matrices of groups, M_{G1} and M_{G2}, resulting from the application of the maximization rule

$$M_{G1} = \begin{pmatrix} (a_{1,g1}; b_{1,g1}) \\ (a_{2,g1}; b_{2,g1}) \\ (a_{3,g1}; b_{3,g1}) \\ (a_{4,g1}; b_{4,g1}) \\ (a_{5,g1}; b_{5,g1}) \end{pmatrix} = \begin{pmatrix} \rho_{1,g1} \\ \rho_{2,g1} \\ \rho_{3,g1} \\ \rho_{4,g1} \\ \rho_{5,g1} \end{pmatrix} \qquad M_{G2} = \begin{pmatrix} (a_{1,g2}; b_{1,g2}) \\ (a_{2,g2}; b_{2,g2}) \\ (a_{3,g2}; b_{3,g2}) \\ (a_{4,g2}; b_{4,g2}) \\ (a_{5,g2}; b_{5,g2}) \end{pmatrix} = \begin{pmatrix} \rho_{1,g2} \\ \rho_{2,g2} \\ \rho_{3,g2} \\ \rho_{4,g2} \\ \rho_{5,g2} \end{pmatrix}$$

$$\textbf{MIN}\{[G_1], [G_2\}, \ldots, [G_h], \ldots, [G_p]\} \text{ or}$$
$$\textbf{MIN}\{(a_{i,g1}; b_{i,g1}); (a_{i,g2}; b_{i,g2}), \ldots, (a_{i,gh}; b_{i,gh}), \ldots, (a_{i,gp}; b_{i,gp})\}$$

from which we obtain for each factor F_i the resultant annotation $(a_{i,R}; b_{i,R})$, in which

$$a_{i,R} = \min\{a_{i,g1}, a_{i,g2}, \ldots, a_{i,gh}, \ldots, a_{i,gp}\}; \text{ and}$$
$$b_{i,R} = \max\{b_{i,g1}, b_{i,g2}, \ldots, b_{i,gh}, \ldots, b_{i,gp}\}.$$

As there are n factors, these results will constitute a column matrix with n lines, which will be called resulting matrix, $M_R = [(a_{i,R}; b_{i,R})] = [\omega_{i,R}]$.

Returning to the example of n = 5 factors, s = 3 sections and **m** = 4 specialists, the application of the maximization rule would be reduced to **MIN** $\{[G_1], [G_2]\}$, obtaining:

$$a_{1,R} = \min\{a_{1,g1}; a_{1,g2}\} \quad \text{and} \quad b_{1,R} = \max\{b_{1,g1}; b_{1,g2}\};$$
$$a_{2,R} = \min\{a_{2,g1}; a_{2,g1}\} \quad \text{and} \quad b_{2,R} = \max\{b_{2,g1}; b_{2,g2}\};$$
$$a_{3,R} = \min\{a_{3,g1}; a_{3,g1}\} \quad \text{and} \quad b_{3,R} = \max\{b_{3,g1}; b_{3,g2}\};$$
$$a_{4,R} = \min\{a_{4,g1}; a_{4,g1}\} \quad \text{and} \quad b_{4,R} = \max\{b_{4,g1}; b_{4,g2}\};$$
$$a_{5,R} = \min\{a_{5,g1}; a_{5,g1}\} \quad \text{and} \quad b_{5,R} = \max\{b_{5,g1}; b_{5,g2}\}.$$

The resulting matrix, column matrix of 5 lines, is represented as in Table 4.5.

The application of the maximization (**MAX**) and minimization (**MIN**) rules, in view of this example that is being analyzed, may be summarized this way:

$$\textbf{MIN}\{\textbf{MAX}[(E_1), (E_4)], \textbf{MAX}[(E_2), (E_3)]\} \quad \text{or} \quad \textbf{MIN}\{[G_1], [G_2]\}.$$

In the applications, the matrices seen in Tables 4.2 and 4.3 (matrices of the weights, researched and researched data), 4.4 (matrices of the groups) and 4.5 (resulting matrix) will be placed as columns in the calculation table, which has the format of Tables 4.6 and 4.7.

Table 4.5 Resultant matrix, (M_R)

$$
M_R = \begin{pmatrix} (a_{1,R};\, b_{1,R}) \\ (a_{2,R};\, b_{2,R}) \\ (a_{3,R};\, b_{3,R}) \\ (a_{4,R};\, b_{4,R}) \\ (a_{5,R};\, b_{5,R}) \end{pmatrix} = \begin{pmatrix} \omega_{1,R} \\ \omega_{2,R} \\ \omega_{3,R} \\ \omega_{4,R} \\ \omega_{5,R} \end{pmatrix}
$$

Table 4.6 Calculation table, with the indication of the bi-valued annotations

F^i	M_P	M_{pq}	M_{Dpq}				M_{G1}	M_{G2}	M_R
	P^i	S_{pi}	E_1	E_2	E_3	E_4	MAX $[E_1, E_4]$	MAX $[E_2, E_3]$	MIN $[G_1, G_2]$
F1	P1	Sp1	$\lambda_{1,1}$	$\lambda_{1,2}$	$\lambda_{1,3}$	$\lambda_{1,4}$	$\rho_{1,g1}$	$\rho_{1,g2}$	$\omega_{1,R}$
F2	P2	Sp2	$\lambda_{2,1}$	$\lambda_{2,2}$	$\lambda_{2,3}$	$\lambda_{2,4}$	$\rho_{2,g1}$	$\rho_{2,g2}$	$\omega_{2,R}$
F3	P3	Sp3	$\lambda_{3,1}$	$\lambda_{3,2}$	$\lambda_{3,3}$	$\lambda_{3,4}$	$\rho_{3,g1}$	$\rho_{3,g2}$	$\omega_{3,R}$
F4	P4	Sp4	$\lambda_{4,1}$	$\lambda_{4,2}$	$\lambda_{4,3}$	$\lambda_{4,4}$	$\rho_{4,g1}$	$\rho_{4,g2}$	$\omega_{4,R}$
F5	P5	Sp5	$\lambda_{5,1}$	$\lambda_{5,2}$	$\lambda_{5,3}$	$\lambda_{5,4}$	$\rho_{5,g1}$	$\rho_{5,g2}$	$\omega_{5,R}$

Table 4.7 Calculation table with indication of the degrees of evidence favorable (**a**) and contrary (**b**) values

F^i	M_P	M_{pq}	M_{Dpq}				M_{G1}	M_{G2}	M_R
	P^i	S_{pi}	E_1	E_2	E_3	E_4	MAX $[E_1, E_4]$	MAX $[E_2, E_3]$	MIN $[G_1, G_2]$
F1	P1	Sp1	$(a_{1,1};\, b_{1,1})$	$(a_{1,2};\, b_{1,2})$	$(a_{1,3};\, b_{1,3})$	$(a_{1,4};\, b_{1,4})$	$(a_{1,g1};\, b_{1,g1})$	$(a_{1,g2};\, b_{1,g2})$	$(a_{1,R};\, b_{1,R})$
F2	P2	Sp2	$(a_{2,1};\, b_{2,1})$	$(a_{2,2};\, b_{2,2})$	$(a_{2,3};\, b_{2,3})$	$(a_{2,4};\, b_{2,4})$	$(a_{2,g1};\, b_{2,g1})$	$(a_{2,g2};\, b_{2,g2})$	$(a_{2,R};\, b_{2,R})$
F3	P3	Sp3	$(a_{3,1};\, b_{3,1})$	$(a_{3,2};\, b_{3,2})$	$(a_{3,3};\, b_{3,3})$	$(a_{3,4};\, b_{3,4})$	$(a_{3,g1};\, b_{3,g1})$	$(a_{3,g2};\, b_{3,g2})$	$(a_{3,R};\, b_{3,R})$
F4	P4	Sp4	$(a_{4,1};\, b_{4,1})$	$(a_{4,2};\, b_{4,2})$	$(a_{4,3};\, b_{4,3})$	$(a_{4,4};\, b_{4,4})$	$(a_{4,g1};\, b_{4,g1})$	$(a_{4,g2};\, b_{4,g2})$	$(a_{4,R};\, b_{4,R})$
F5	P5	Sp5	$(a_{5,1};\, b_{5,1})$	$(a_{5,2};\, b_{5,2})$	$(a_{5,3};\, b_{5,3})$	$(a_{5,4};\, b_{5,4})$	$(a_{5,g1};\, b_{5,g1})$	$(a_{5,g2};\, b_{5,g2})$	$(a_{5,R};\, b_{5,R})$

The values of the resulting favorable ($a_{i,R}$) and contrary ($b_{i,R}$) evidence, obtained for all the factors, enable us to determine the influence of each factor in the feasibility of the enterprise. That is performed through the para-analyzing algorithm.

Just plot the pairs ($a_{i,R}$; $b_{i,R}$) in the cartesian plane, obtaining n points that represent the n factors, and verify the position of these points in lattice τ. If the point belongs to the truth region, the corresponding factor influences in the sense of recommending the execution of the enterprise; if it belongs to the falsity region, the factor recommends the non-execution of the enterprise; but if the point belongs to a different region from these ones, it is said that the factor is inconclusive, that is, that neither the execution nor the non-execution of the enterprise are recommended.

This analysis of the influence of the factors may also be performed calculating the degree of certainty ($H_i = a_{i,R} - b_{i,R}$) for each factor and applying the decision rule. If $H_i \geq NE$, factor F_i recommends the execution of the enterprise; if $H_i \leq -NE$, factor F_i recommends the non-execution of the enterprise; and if $-NE < H_i < NE$, factor F_i is inconclusive, that is, does not recommend the execution nor the non-execution of the enterprise.

4.2.7 Determination of the Center of Gravity

Generally, there is not much interest in knowing the influence of each factor, isolatedly. However, it is fundamentally important to know the joint influence of all the factors on the feasibility of the enterprise, as it leads to the final decision. The joint influence of the factors is determined by the analysis of the center of gravity (W) of the points that represent them in the cartesian plane (in lattice τ). To determine the center of gravity, its coordinates are calculated, which are the degrees of favorable (a_w) and contrary (b_w) evidences. The degree of favorable evidence of the center of gravity (a_w) is equal to the weighted average of the favorable evidences resulting from all the factors, taking as coefficient (weights) the weights (P_i) attributed to the factors by the specialists. Analogously, the degree of contrary evidence of the center of gravity (b_w) is calculated.

$$a_W = \frac{\sum_{i=1}^{n} P_i a_{i,R}}{\sum_{i=1}^{n} P_i} \quad \text{and} \quad b_W = \frac{\sum_{i=1}^{n} P_i b_{i,R}}{\sum_{i=1}^{n} P_i} \tag{4.4}$$

In the specific case in which all the factors have equal weights (P_1), the weighted averages at (4.4) transform into arithmetic averages and the center of gravity of the points that represent the factors become the geometrical center of these points. In this case:

$$a_W = \frac{\sum_{i=1}^{n} a_{i,R}}{n} \quad \text{and} \quad b_W = \frac{\sum_{i=1}^{n} b_{i,R}}{n} \tag{4.5}$$

4.2.8 Decision Making

After determining the values of favorable (a_w) and contrary (b_w) evidences of the center of gravity, it is already possible to perform the final decision making, through the para-analyzing algorithm.

For this purpose, just plot the ordered pair (a_w; b_w) in the cartesian plane and verify which region of lattice τ the center of gravity **W** belongs to. If it belongs to the truth region, the decision is favorable, that is, the enterprise is feasible; if it belongs to the falsity region, the decision is unfavorable, that is, the enterprise is not feasible; but if it belongs to any region of lattice τ other than these two, it is said that is the analysis is inconclusive. In this case, no conclusion is made for the feasibility of the enterprise, nor for its unfeasibility; we just say that the analysis was not conclusive and that, if it is of interest, new studies must be conducted to try to reach to a positive conclusion (feasibility or unfeasibility), always having the corresponding non-extreme states as support.

Another way to obtain the final decision is by the application of the decision rule. In this case, just calculate the degree of certainty of the center of gravity ($H_w = a_w - b_w$) and apply the decision rule. If $H_w \geq NE$, the decision is favorable and the result recommends the execution of the enterprise (feasible); if $H_w \leq -NE$, the decision is unfavorable and it recommends the non-execution of the enterprise; and if $-NE < H_w < NE$, it is said that the analysis is inconclusive, that is, the result does not recommend the execution nor the non-execution of the enterprise. It only suggests that, if it is of interest, new studies are conducted to try solving the indecision.

It must be observed, then, that the degree of certainty of the center of gravity (H_w) is the final, well-determined number, which enables the sought decision. The whole process ends up leading to this number, essential for the decision making with the obtention of new evidences.

To close this chapter, it is worth to notice that all the described operations— search for the opinions of the specialists in the database, once the research result is known (stage 5); calculation of the values of the favorable and contrary evidences for each one of the factors (stage 6); calculation of the values of the favorable and contrary evidences of the center of gravity (stage 7); and the decision making (stage 8)—are performed by a computer program developed with the aid of the Excel electronic spreadsheet. This program will be called Calculation Program For The Paraconsistent Decision Method (CP of the PDM).

With the intention of aiding the ones less trained in the use of the Excel spreadsheet, Chap. 5 shows how to perform the assembly of the CP of the PDM.

Chapter 5
Calculation Program
for the Paraconsistent Decision Method
(PDM'S CP)

The Calculation Program of the Paraconsistent Decision-Making (PDM's CP) is an application of the Excel spreadsheet. This way, with the intention to help the ones less familiar with this worksheet, we took the liberty to show how, with the aid of Excel, the necessary calculations are made for the PDM application. Beforehand, we ask for the comprehension of the Excel experts, so that they disregard if something has not been done the best way, and we request their kindness to present the suggestions of improvements they deem pertinent.

Of course, these calculations could be done without the computing resources, but they would become so time-consuming and tedious that it would impair the method.

To make this exhibition tuned with the method presentation text (Chap. 4), it will follow the sequence of the PDM stages, which are executed by the calculation program. We will use the Office 2003 Excel spreadsheet; for version 2007, some small adaptations will be necessary.

5.1 Search for the Expert's Opinions on the Database, Once the Research Result Is Known (Stage 5)

To make the explanation more accessible, it will be accompanied by a numeric example. For this purpose, the following database will be considered.

That is, a database corresponding to analysis of four influence factors, three sections for each factor and four specialists will be utilized.

To the reader: if you wish to follow the explanations and, at the same time, execute them in an Excel spreadsheet, it will be easier if you keep the numbering of the lines and the columns used in this text.

Suppose that the research result has accused that the factors F_1 to F_4 are in the conditions translated by sections S_2, S_1 S_2 and S_3, respectively. Therefore, these are the elements of the researched matrix. Thus, the transpose of this matrix is

© Springer International Publishing AG, part of Springer Nature 2018
F. R. de Carvalho and J. M. Abe, *A Paraconsistent Decision-Making Method*,
Smart Innovation, Systems and Technologies 87,
https://doi.org/10.1007/978-3-319-74110-9_5

Table 5.1 Data base for four factors F_i, three sections S_j and four specialists E_k

	A	B	C	D	E	F	G	H	I	J	K	L
1	Data base											
2												
3	Factor	Weight	Section	F and S	Spec 1		Spec 2		Spec 3		Spec 4	
4					$a_{i,j,1}$	$b_{i,j,1}$	$a_{i,j,2}$	$b_{i,j,2}$	$a_{i,j,3}$	$b_{i,j,3}$	$a_{i,j,4}$	$b_{i,j,4}$
5	F1	4	S1	F1S1	1.0	0.0	0.9	0.1	1.0	0.2	0.8	0.3
6			S2	F1S2	**0.7**	**0.6**	**0.6**	**0.4**	**0.6**	**0.6**	**0.5**	**0.6**
7			S3	F1S3	0.3	1.0	0.3	1.0	0.2	0.8	0.2	1.0
8	F2	3	S1	F2S1	**0.9**	**0.2**	**1.0**	**0.2**	**0.9**	**0.1**	**0.8**	**0.1**
9			S2	F2S2	0.6	0.5	0.6	0.6	0.4	0.4	0.5	0.4
10			S3	F2S3	0.3	0.9	0.2	0.8	0.1	0.8	0.0	0.9
11	F3	2	S1	F3S1	0.9	0.2	0.8	0.2	0.8	0.0	0.7	0.2
12			S2	F3S2	**0.5**	**0.4**	**0.4**	**0.4**	**0.6**	**0.5**	**0.5**	**0.5**
13			S3	F3S3	0.3	1.0	0,0	1.0	0.3	1.0	0.1	0.9
14	F4	1	S1	F4S1	1.0	0.2	0.8	0.0	1.0	0.2	0.9	0.4
15			S2	F4S2	0.5	0.6	0.6	0.6	0.6	0.4	0.5	0.6
16			S3	F4S3	**0.1**	**0.9**	**0.2**	**0.9**	**0.2**	**1.0**	**0.0**	**0.9**

$[S_{pj}]^t = [S_2, S_1, S_2, S_3]$. It will be considered that the transpose of the matrix of the weights is $[P_i]^t = [4, 3, 2, 1]$, which is already contained in the database.

The problem is to seek in the database the values of the degrees of evidence, favorable and contrary, corresponding to these sections, which are highlighted in bold in Table 5.1.

This task is resolved with the aid of the PROCV function, combined with the CONCATENATE (concatenate) function. The first one has three arguments: Valor_procurado (Sought value), Matriz_tabela (Table matrix) and Número_índice coluna (Column index number); the second one has as many arguments as the texts to be concatenated (text 1, text 2, …). In the CP of the PDM we will have, typically, only two texts to be concatenated: the indicator of the factor (F_i) and the indicator of the section (S_j).

This, the combination of the two functions results in the function:

PROCV(CONCATENAR (Fator; Seção); Matriz_tabela; Número_índice_ coluna).

For the application of this function, it is required to insert one more column, D, in the database (key or base column), where the two texts of the CONCATENAR function (Factor and Section) are grouped (see Table 5.1). This way, the Matriz_tabela starts being the segment of the database that goes from line 5 to 16 and from column D to L, that is, ($D$5:$L$16). Therefore, column D is column 1 of the Matriz_tabela.

The order of the column (Número_índice_coluna), as the name itself says, is the order of the column of the Table matrix where the data you wish to search is. If you want the degree of favorable evidence of specialist 1 ($a_{i,j,1}$), the column is E and the order of the column is 2, because the first column of the Matriz_tabela is D; if you want the degree of contrary evidence os specialist 1 ($b_{i,j,1}$), the column is F and the order of the column is 3, because the first column of the Matriz_tabela is D; and so on, until you reach column L, whose order is 9.

One detail: for the Excel to execute the PROCV function, it is required that the base column (or key) of the Matriz_tabela (column D) be placed in ascending order. Here, in the example that is being analyzed, column D is already ordered. We must not forget, when placing the base column in ascending order, to make all the other columns follow the order.

Thus, the first part of the calculation table of the CP of the PDM will have the form presented in Table 5.3, in which the 4th column (P) will contain the values of the degrees of favorable evidence attributed by specialist E_1 to the factors, in the conditions of the researched sections ($a_{i,j,1}$); the 5th column (Q), the degrees of contrary evidence ($b_{i,j,1}$); and so forth, until the last column (W), which will contain the degrees of contrary evidence attributed by specialist E_4. In Table 5.2 are the formulas that enable the search for these values in the database (Table 5.1).

Observe that, in Table 5.2, there are only the formulas corresponding to columns P and W, because the other ones (columns Q, R, S, T, U and V) are obtained analogously, just exchanging the column order (Número_índice_coluna) for 3, 4, 5, 6, 7 and 8, respectively.

Applying these 32 formulas (4 lines and 8 columns), we obtain Table 5.3, which contains precisely the values highlighted in bold in the database (Table 5.1).

5.2   Obtention of the Resulting Values from the Favorable Evidence and the Contrary Evidence for Each One of the Factors (Stage 6)

Let's suppose the four experts are distributed into two groups: Group A, constituted by specialists E_1 and E_2, and Group B, constituted by specialists E_3 and E_4.

The maximization (**MAX** operator) is applied intragroup, that is, inside group A and group B, and the minimization (**MIN** operator) is applied intergroup, that is, among the results obtained by the maximization applied to groups A and B. Thus, the scheme to be adopted to apply the techniques of logic E_T is the following:

$$\textbf{MIN}\{\textbf{MAX}[(E_1), (E_2)]; \textbf{MAX}[(E_3), (E_4)]\} \quad \text{or}$$
$$\textbf{MIN}\{[G_A]; [G_B]\}.$$

Therefore, to obtain the values of favorable evidence and contrary evidence for each one of the four factors, the formulas to be applied are the ones of Table 5.4, which is the continuation of the PDM calculation table.

Applying these formulas to the values in Table 5.3, we obtain Table 5.5.

Table 5.2 Formulas to search the data in data base

	M	N	O	P	Q	R	S	T	U	V	W
1	Calculation table										
2											
3	Factor	Weight	Section	Spec E1		E2		E3		Spec E4	
4				$a_{i,1}$	$a_{i,1}$	$a_{i,2}$	$b_{i,2}$	$a_{i,3}$	$b_{i,3}$	$a_{i,4}$	$b_{i,4}$
5	F1	P1	S2	'=PROCV(CONCATENAR ($M5;$N5);$D$5:$L$16;2)							'=PROCV(CONCATENAR ($M5;$N5);$D$5:$L$16;9)
6	F2	P2	S1	'=PROCV(CONCATENAR ($M6;$N6);$D$5:$L$16;2)							'=PROCV(CONCATENAR ($M6;$N6);$D$5:$L$16;9)
7	F3	P3	S2	'=PROCV(CONCATENAR ($M7;$N7);$D$5:$L$16;2)							'=PROCV(CONCATENAR ($M7;$N7);$D$5:$L$16;9)
8	F4	P4	S3	'=PROCV(CONCATENAR ($M8;$N8);$D$5:$L$16;2)							'=PROCV(CONCATENAR ($M8;$N8);$D$5:$L$16;9)
9	'=SOMA(N5:N8)										

Table 5.3 Values found in data base by application of the formulas of Table 5.2

	M	N	O	P	Q	R	S	T	U	V	W
1	Calculation table										
2											
3	Factor	Weight	Section	Spec E1		Spec E2		Spec E3		Spec E4	
4				$a_{i,1}$	$b_{i,1}$	$a_{i,2}$	$b_{i,2}$	$a_{i,3}$	$b_{i,3}$	$a_{i,4}$	$b_{i,4}$
5	F1	4	S2	0.7	0.6	0.6	0.4	0.6	0.6	0.5	0.6
6	F2	3	S1	0.9	0.2	1.0	0.2	0.9	0.1	0.8	0.1
7	F3	2	S2	0.5	0.4	0.4	0.4	0.6	0.5	0.5	0.5
8	F4	1	S3	0.1	0.9	0.2	0.9	0.2	1.0	0.0	0.9
9		10									

Table 5.4 Formulas to apply the maximization (MAX) and minimization (MIN) rules

	X	Y	Z	AA	AB	AC	AD	AE
1	Calculation table							
2	Group A		Group B		Resultant degrees		Certainty and	
3	**MAX [E1, E2]**		**MAX [E3, E4]**		**MIN {GA, GB}**		contradiction	
4	$a_{i,A}$	$b_{i,A}$	$a_{i,j,B}$	$b_{i,B}$	$a_{i,R}$	$b_{i,R}$	**H**	**G**
5	'=MAXIMUM (P5;R5)	'=MINIMUM (Q5;S5)			'=MINIMUM (X5;Z5)	'=MAXIMUM (Y5;AA5)	'=AB5 −AC5	'=AB5 +AC5−1
6	'=MAXIMUM (P6;R6)	'=MINIMUM (Q6;S6)			'=MINIMUM (X6;Z6)	'=MAXIMUM (Y6;AA6)	'=AB6 −AC6	'=AB6 +AC6−1
7	'=MAXIMUM (P7;R7)	'=MINIMUM (Q7;S7)			'=MINIMUM (X7;Z7)	'=MAXIMUM (Y7;AA7)	'=AB7 −AC7	'=AB7 +AC7−1
8	'=MAXIMUM (P8;R8)	'=MINIMUM (Q8;S8)			'=MINIMUM (X8;Z8)	'=MAXIMUM (Y8;AA8)	'=AB8 −AC8	'=AB8 +AC8−1
9	**Baricenter W: average of resultant degrees**				'=AG9/N9	'=AH9/N9	'=AB9 −AC9	'=AB9 +AC9−1

5.3 Calculation of the Values of the Degrees of Evidence, Favorable and Contrary, of the Center of Gravity (Stage 7)

This stage was already solved in Tables 5.4 and 5.5, line 9, columns AB and AC.

The formulas utilized for the calculation of the degrees of evidence of the center of gravity are the ones of weighted average:

Table 5.5 Results of application of the maximization (MAX) and minimization (MIN) rules

	X	Y	Z	AA	AB	AC	AD	AE
1	Calculation table							
2	Group A		Group B		Resultant degrees		Certainty and contradiction	
3	MAX [E1, E2]		MAX [E3, E4]		MIN {GA, GB}			
4	$a_{i,A}$	$b_{i,A}$	$a_{i,B}$	$b_{i,A}$	$a_{i,R}$	$b_{i,R}$	H	G
5	0.7	0.4	0.6	0.6	0.6	0.6	0.0	0.2
6	1.0	0.2	0.9	0.1	0.9	0.2	0.7	0.1
7	0.5	0.4	0.6	0.5	0.5	0.5	0.0	0.0
8	0.2	0.9	0.2	0.9	0.2	0.9	−0.7	0.1
9	**Baricenter W: average of resultant degrees**				**0.63**	**0.49**	**0.14**	**0.12**

$$a_W = \frac{\sum_{i=1}^{n} P_i a_{i,R}}{\sum_{i=1}^{n} P_i} \quad \text{and} \quad b_W = \frac{\sum_{i=1}^{n} P_i b_{i,R}}{\sum_{i=1}^{n} P_i} \tag{4.4}$$

The weighting that appears in the numerators of these formulas is shown in Tables 5.6 and 5.7, columns AG and AH, and the sum of the weights of the denominators is obtained in cell N9 of Table 5.2.

Observe that, in the particular case of the weights being all equal, the weighted averages above are reduced to the particular case of the arithmetic averages translated by the formulas:

$$a_W = \frac{\sum_{i=1}^{n} a_{i,R}}{n} \quad \text{and} \quad b_W = \frac{\sum_{i=1}^{n} b_{i,R}}{n} \tag{4.5}$$

However, this does not alter the calculation at all, as the CP of the PDM, as it is being constructed here, already contemplates this particularization.

5.4 The Decision Making (Stage 8)

The decision making is performed calculating the degree of certainty and applying the decision rule. This may be done for each factor isolatedly ($H_i = a_{i,R} - b_{i,R}$), when you wish to know the influence of each factor on the enterprise; or for the center of gravity ($H_w = a_w - b_w$), when you wish to know the joint influence of all the factors on the enterprise (Column AD of Table 5.3).

To know if the data related to each factor are contradictory or not, we calculate the degree of uncertainty, $G_i = a_{i,R} + b_{i,R} - 1$ (Column AE of Table 5.3).

Table 5.6 Formulas to decision-making and to do the weighting in the calculation of the resultant degrees averages

	AF	AG	AH
1	Calculation table		
2	Level of requirement LR	Weighting of the resultant	
3		degrees	
4	Decision	$P_i \times a_{i,R}$	$P_i \times b_{i,R}$
5	'=IF(AD5 <=−AF$3; "UNVIABLE";IF(AD5<AF$3; "NOT CONCLUSIVE";"VIABLE"))	N5*AB5	N5*AC5
6	'=IF(AD6<=−AF$3; "UNVIABLE";IF(AD6<AF$3; "NOT CONCLUSIVE";"VIABLE"))	N6*AB6	N6*AC6
7	'=IF(AD7<=−AF$3; "UNVIABLE";IF(AD7<AF$3; "NOT CONCLUSIVE";"VIABLE"))	N7*AB7	N7*AC7
8	'=IF(AD8<=−AF$3; "UNVIABLE";IF(AD8<AF$3; "NOT CONCLUSIVE";"VIABLE"))	N8*AB8	N8*AC8
9	**'=IF(AD9<=−AF$3;"UNVIABLE";IF(AD9<AF$3; "NOT CONCLUSIVE";"VIABLE"))**	'=SUM (AG5:AG8)	'=SUM (AH5:AH8)

Table 5.7 Decision-making and weighting of the resultant degrees

	M	AD	AE	AF	AG	AH
1	Calculation table					
2	Factor	Certainty and		Level of requirement **0.60**	Weighting of the	
3		contradiction			resultant degrees	
4		**H**	**G**	Decision	$P_i \times a_{i,R}$	$P_i \times b_{i,R}$
5	F1	0.0	0.2	Not conclusive	2.4	2.4
6	F2	0.7	0.3	Viable	2.7	0.6
7	F3	0.0	0.0	Not conclusive	1.0	1.0
8	F4	−0.7	0.1	Unviable	0.2	0.9
9	Baricenter	**0.14**	**0.12**	**Not conclusive**	**6.3**	**4.9**

To apply the decision rule and make the weighting of the resultant degrees of the factors for the calculation of the degrees of evidence of the center of gravity, the formulas are the ones in Table 5.6.

The Table 5.8 allows one complete view of PDM calculation table.

The columns of this table vary according to the quantity of specialists, the way to form the groups, etc. The quantity of lines depends only on the number of considered factors.

Table 5.8 Complete table of calculations of the Paraconsistent Decision-Making Method (PDM)

Factor	Weight	Section	Spec E1 $a_{i,1}$	$b_{i,1}$	E2 $a_{i,2}$	$b_{i,2}$	E3 $a_{i,3}$	$b_{i,3}$	Spec E4 $a_{i,4}$	$b_{i,4}$	Group A MAX[E1,E2] $a_{i,A}$	$b_{i,A}$	Group B MAX[E3,E4] $a_{i,B}$	$b_{i,B}$	Resultant degrees MIN{GA,GB} $a_{i,LR}$	$b_{i,LR}$	Certainty and Contradiction H	G	Level of requirement 0.60 Decision	Weighting of resultant degrees $P_i \times a_{i,LR}$	$P_i \times b_{i,LR}$
F1	4	S2	0.7	0.6	0.6	0.4	0.6	0.6	0.5	0.6	0.7	0.4	0.6	0.6	0.6	0.6	0.0	0.2	Not conclusive	2.4	2.4
F2	3	S1	0.9	0.2	1.0	0.2	0.9	0.1	0.8	0.1	1.0	0.2	0.9	0.1	0.9	0.2	0.7	0.1	Viable	2.7	0.6
F3	2	S2	0.5	0.4	0.4	0.4	0.6	0.5	0.5	0.5	0.5	0.4	0.6	0.5	0.5	0.5	0.0	0.0	Not conclusive	1.0	1.0
F4	1	S3	0.1	0.9	0.2	0.9	0.2	1.0	0.0	0.9	0.2	0.9	0.2	0.9	0.2	0.9	-0.7	0.1	Unviable	0.2	0.9
	10														0.63	0.49	0.14	0.12	Not conclusive	6.3	4.9

Baricenter W: average of resultant degrees

5.5   The Construction of the Para-Analyzer Algorithm (Chart)

To meet some people's curiosity, below we show how to utilize Excel to assemble the para-analyzer algorithm. That is, to make the chart of the degree of contrary evidence *versus* the degree of favorable evidence with the limit lines and the region borders, and plot in it the representative points of the factors and the center of gravity.

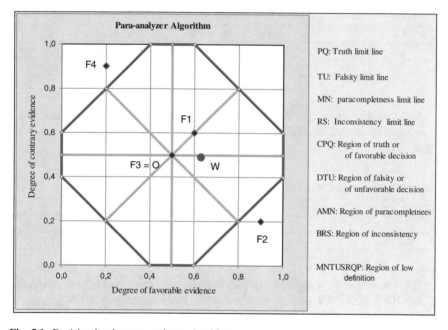

Fig. 5.1 Decision by the para-analyzer algorithm

One idea is exposed below (we say one idea because there are several versions of Excel and, from one to the other, the procedures vary slightly; we are taking Excel 2003 as a base).

Firstly, the limit lines and the borders may be made. For this purpose, just assemble two sequences of points (see Fig. 5.1): MPQRSUTNM, which defines the limit lines and the decision regions, and EOFOGOHOIOJOKOL, which defines the borders between the inconclusive regions.

The coordinates (degrees of evidence) of the points of sequence MPQRSUTNM depend exclusively on the requirement level (NE) adopted. This way, they may be placed according to this parameter, obtaining Table 5.9.

Table 5.9 Coordinates of the limit lines extrems in function of the level of requirement

Dot	M	P	Q	R	S	U	T	N	M
a	1 − NE	NE	1.0	1.0	NE	1 − NE	0.0	0.0	1 − NE
b	0.0	0.0	1 − NE	NE	1.0	1.0	NE	1 − NE	0.0

This way, as in the example we are analyzing, the requirement level is in cell AF3, replacing NE for AF3, Table 5.9 is modified to Table 5.10.

Table 5.10 Coordinates of the limit lines extrems in function of the level of requirement in AF3

Dot	M	P	Q	R	S	U	T	N	M
a	1 − AF3	AF3	1.0	1.0	AF3	1 − AF3	0.0	0.0	1 − AF3
b	0.0	0.0	1 − AF3	AF3	1.0	1.0	AF3	1 − AF3	0.0

Also considering that, in the example in analysis, the adopted requirement level is equal to 0.60, and placing this value in cell AF3, Excel calculates the values of Table 5.11.

Table 5.11 Coordinates of the limit lines extrems for the level of requirement 0.60

Dot	M	P	Q	R	S	U	T	N	M
a	0.4	0.6	1.0	1.0	0.6	0,4	0.0	0.0	0.4
b	0.0	0.0	0.4	0.6	1.0	1.0	0.6	0.4	0.0

The coordinates (degrees of evidence) of the points of sequence EOFOGOHOIOJOKOL also depend on the NE. Placed according to this parameter, we obtain Table 5.12.

Table 5.12 Coordinates of the frontier extrems in function of the level of requirement

Dot	O	E	F	G	H	I	J	K	L
a	0.5	(1 − NE)/2	(1 + NE)/2	(1 + NE)/2	(1 − NE)/2	0.0	1.0	0.5	0.5
b	0.5	(1 − NE)/2	(1 + NE)/2	(1 − NE)/2	(1 + NE)/2	0.5	0.5	0.0	1.0

For the requirement level NE in AF3, we obtain Table 5.13.

Table 5.13 Coordinates of the frontier extrems in function of the level of requirement in AF3

Dot	O	E	F	G	H	I	J	K	L
a	0.5	(1 − AF3)/2	(1 + AF3)/2	(1 + AF3)/2	(1-AF3)/2	0.0	1.0	0.5	0.5
b	0.5	(1 − AF3)/2	(1 + AF3)/2	(1 − AF3)/2	(1 + AF3)/2	0.5	0.5	0.0	1.0

Placing the value 0.60 of the requirement level in cell AF3, Excel calculates the values of Table 5.14.

Table 5.14 Coordinates of the frontier extrems for the level of requirement 0.60

Dot	O	E	F	G	H	I	J	K	L
a	0.5	0.2	0.8	0.8	0.2	0.0	1.0	0.5	0.5
b	0.5	0.2	0.8	0.2	0.8	0.5	0.5	0.0	1.0

With these values, Excel makes the para-analyzer algorithm (the chart) and alters it every time the requirement level (cell AF3, in the case of this example) is altered. The operations so that it occurs are seen further ahead.

To represent the points that translate the factors in the para-analyzer algorithm, just represent the sequence of points $F_i = (a_{i,R}; b_{i,R})$, which, in the example that is being considered, is shown in Table 5.15.

Table 5.15 Resultant evidence degrees of the factors

Factor	F_1	F_2	F_3	F_4	Factor	F_1	F_2	F_3	F_4
$a_{i,R}$	AB5	AB6	AB7	AB8	$a_{i,R}$	0.6	0.9	0.5	0.2
$b_{i,R}$	AC5	AC6	AC7	AC8	$b_{i,R}$	0.6	0.1	0.4	0.9

To locate the center of gravity W, just represent the sequence $W = (a_{i,w}; b_{i,w})$ constituted by one single point, the center of gravity. In the example: $W = (AB9; AC9) = (0.63; 0.44)$ (Table 5.16).

Table 5.16 Resultant evidence degrees of the baricenter

Baricenter		Baricenter	
$a_{i,w}$	AB9	$a_{i,w}$	0.63
$b_{i,w}$	AC9	$b_{i,w}$	0.44

Still thinking about the reader who is not so familiar with Excel (and we count with the experts' comprehension in this spreadsheet), a detailing of the sequence of necessary steps for the assembly of the para-analyzer algorithm (of the chart) will be done.

1st Step: in columns from AL to AR (these seven columns were established to facilitate the exposition, but it could be any other seven), we make a table, placing: in AL, from line 4 until line 27, the sequences MPQRSUTNM, which defines the limit lines and the decision regions, and EOFOGOHOIOJOKOL, which defines the borders between the inconclusive regions;

(a) in AM, the values of *a*, as contained in Tables 5.9 and 5.12;
(b) in AN, the values of *b*, as contained in Tables 5.9 and 5.12;
(c) in AO, the values of *a*, as contained in Tables 5.10 and 5.13;

(d) in AP, the values of **b**, as contained in Tables 5.10 and 5.13;
(e) in AQ, the values of **a**, as contained in Tables 5.11and 5.14;
(f) in AR, the values of **b**, as contained in Tables 5.11 and e 5.14.

The reader must have already realized that from these six items, only items (e) and (f) are necessary, as long as the formulas indicated in items (c) and (d) are used from the start. The intention here was to show the way, item by item, for the ones less trained in the Excel spreadsheet.

With that, we obtain Table 5.17, which exhibits the formulas and calculations already executed (always remembering that in the example in the analysis, the NE is in cell AF3 and is worth 0.60).

Table 5.17 Values that result in the limit lines and the borders of the low definition regions

Limit lines e Frontiers of the low definition regions						
Point	Formulas		Formulas		Values for LR = 0.60	
	a	*b*	*a*	*b*	*a*	*b*
M	1 − NE	0.0	1 − AF3	0.0	0.4	0.0
P	NE	0.0	AF3	0.0	0.6	0.0
Q	1.0	1 − NE	1.0	1 − AF3	1.0	0.4
R	1.0	NE	1.0	AF3	1.0	0.6
S	NE	1.0	AF3	1.0	0.6	1.0
U	1 − NE	1.0	1 − AF3	1.0	0.4	1.0
T	0.0	NE	0.0	AF3	0.0	0.6
N	0.0	1 − NE	0.0	1 − AF3	0.0	0.4
M	1 − NE	0.0	1 − AF3	0.0	0.4	0.0
E	(1 − NE)/2	(1 − NE)/2	(1 − AF3)/2	(1 − AF3)/2	0.2	0.2
O	0.5	0.5	0.5	0.5	0.5	0.5
F	(1 + NE)/2	(1 + NE)/2	(1 + AF3)/2	(1 + AF3)/2	0.8	0.2
O	0.5	0.5	0.5	0.5	0.5	0.5
G	(1 + NE)/2	(1 − NE)/2	(1 + AF3)/2	(1 − AF3)/2	0.8	0.8
O	0.5	0.5	0.5	0.5	0.5	0.5
H	(1 − NE)/2	(1 + NE)/2	(1 − AF3)/2	(1 + AF3)/2	0.2	0.8
O	0.5	0.5	0.5	0.5	0.5	0.5
I	0.0	0.5	0.0	0.5	0.0	0.5
O	0.5	0.5	0.5	0.5	0.5	0.5
J	1.0	0.5	1.0	0.5	1.0	0.5
O	0.5	0.5	0.5	0.5	0.5	0.5
K	0.5	0.0	0.5	0.0	0.5	0.0
O	0.5	0.5	0.5	0.5	0.5	0.5
L	0.5	1.0	0.5	1.0	0.5	1.0

Note CP is the name of the spreadsheet page (it appears at the bottom part of the spreadsheet), in which the CP of the PDM is being assembled. For each following step, click on "Advance" again

2nd Step: the windows "Insert", "Chart", "Dispersion (XY)" (here we choose the third one on the left, with line segments), "Advance", "Series", "Add" are opened, successively.

3rd Step: to make the limit lines:
Name: Limit lines
X-Values: PC!$AQ$4:$AQ$12
Y-Values: PC!$AR$4:$AR$12

4th Step: to make the internal borders of the low definition region:
Name: Borders
X-Values: PC!$AQ$13:$AQ$27
Y-Values: PC!$AR$13:$AR$27

5th Step: represent the points that translate the factors:
Name: Factors
X-Values: PC!$AB$5:$AB$8
Y-Values: PC!$AC$5:$AC$8

6th Step: represent the center of gravity:
Name: Center of gravity
X-Values: PC!$AB$9
Y-Values: PC!$AC$9
To conclude, click on "Advance", "Advance", "Conclude".

From that point on, just format the chart at each person's criteria and taste. For this purpose, almost always, the action is clicking with the right button of the mouse on what you wish to format. However, that will be left for each person to find out, gradually, on their behalf and rhythm.

As it was possible to observe during the presentation of this text, the calculation table and the para-analyzer algorithm are dynamic, that is, they vary with the alteration of the elements that feed them. The main variations to be considered are the researched matrix, constituted by the sections (S_{pj}) obtained in the field research, the matrix of the weights (P_i) and the requirement level (NE).

You, reader, who followed the text of this chapter and assembled the CP of the PDM, will be able to check what you have done in Appendix A—CP of the PDM—Solution for Chap. 5, where the program is assembled exactly as suggested. If you have not, you may use the solution in Appendix A to solve the exercises.

Besides that, there is another appendix, Appendix B—CP of the PDM—Generic Version, which brings the program partially assembled and allows you to make the calculations in different situations.

Exercises

5.1 To feel the changes of the calculation spreadsheet and of the para-analyzer algorithm, make some alterations, (a) In column O, alter the researched matrix to $[S_1, S_1, S_1, S_1,]^t$ and verify the result in the table and in the chart; (b) now, alter the requirement level (cell AF3) to 0.70 and verify the result once more; (c) repeat (b) for NE = 0.85.

5.2 Using the transpose of [S_3, S_3, S_3, S_3,] as researched matrix, calculate the degrees of evidence, favorable and contrary, and the degree of certainty of the center of gravity, and answer what is the decision for the requirement levels (a) 0.60; (b) 0.70 and (c) 0.85.

5.3 Repeat exercise 5.2, using the transpose of [S_1, S_1, S_2, S_3,] as researched matrix, and as requirement level (a) 0.50; (b) 0.65 and (c) 0.85.

5.4 Repeat exercise 5.3, only inverting the order of the factor weights, that is, adopting as transpose of the matrix of the weights: $[P_i]^t = [1, 2, 3, 4]$.

Answers

5.1 (a) Besides the alterations of values, we highlight the following: the degrees of evidence of the center of gravity and its degree of certainty are altered to W = (0.93; 0.20); H_w = 0.73; the decision column (AF) was altered from (NC, V, NC, I and **NC**) to (V, V, V, V and **V**); therefore, the decision that was "Inconclusive" changed to "Feasible"; in the chart, the four points that represent the factors moved to the truth region (one of them, F, coincides with G, on the truth limit line);

(b) The numbers in the table are not altered; the decision column changes to: (V, V, V, NC and **V**); the limit lines change position, so that F_3 leaves the truth region; the decision continued being "Feasible".

(c) The decision column changes to: (NC, NC, NC, NC and **NC**); the limit lines change position, so that F_1 continues in the truth region; the decision changed to "Inconclusive".

5.2 W = (0.19; 0.93); H_w = −0.74;

   (a) (I, I, I, I and **I**); Infeasible;
   (b) (I, I, I, I, and **I**); Infeasible;
   (c) (NC, NC, NC, NC and **NC**); Inconclusive.

5.3 W = (0.79; 0.36) and H_w = 0.46;

   (a) (V, V, NC, I and **NC**); Inconclusive;
   (b) (V, V, NC, I and **NC**); Inconclusive;
   (c) (NC, NC, NC, NC and **NC**); Inconclusive.

5.4 W = (0.51; 0.57) and H_w = −0.06;

   (a) (V, V, NC, I and **NC**); Inconclusive;
   (b) (V, V, NC, I and **NC**); Inconclusive;
   (c) (NC, NC, NC, NC and **NC**); Inconclusive

Chapter 6
Application Examples

In this chapter, several applications of the PDM will be analyzed in detail, trying to introduce a new nuance of the method in each one. Due to this accumulation of derivatives of the method, the applications will become, gradually, more difficult and demand more from the reader.

In the first application (6.1) the utilization of the PDM in a real problem connected to school administration [44] is addressed. In this application, we sought to utilize a simpler situation. A small number of influence factors, all with equal weights, only three sections to translate the conditions of these factors and four specialists. We also sought to analyze the reliability of the method and utilize it so that the reader trains the assembly of the calculation program of the paraconsistent decision method (CP of the PDM). Some exercises are proposed, with the respective answers.

In the following application (6.2), the PDM is utilized to analyze the feasibility of the launch of a new product in the market [45]. Therefore, it is an application of the method in the Marketing area and constitutes a problem with which the professionals in this area constantly come across, besides being extremely important. Here, some additional aspects are addressed. The number of utilized influence factors is ten, five sections were established to translate the conditions of these factors, but the number of specialists continued to be four. In this application, at the end, we sought to show the sensitivity of the analysis in relation to the requirement level, verifying in which way the decision form varies for different values of this parameter.

Application 6.3 is strictly linked to Production Engineering, as it is a typical problem of this Engineering area. This application shows how it is possible to analyze the implementation of a factory project; the analysis is conducted, utilizing

© Springer International Publishing AG, part of Springer Nature 2018
F. R. de Carvalho and J. M. Abe, *A Paraconsistent Decision-Making Method,*
Smart Innovation, Systems and Technologies 87,
https://doi.org/10.1007/978-3-319-74110-9_6

factors that are as close to the reality of this kind of project as possible, and attributing different weights to the factors [46]. In the end, a sensitivity analysis of the decision about the required level is performed.

In application 6.4, the PDM is utilized to analyze a problem discussed by Chalos, in 1922 [22], who seeks to verify if there is an advantage or not in replacing the old manufacturing system with the traditional technology for a modern manufacturing system, provided with advanced technologies [48]. The following modern manufacturing systems are discussed: *CAD/CAM*—Computer-Automated Design and Manufacture; *GT/CM*—Group Technology and Cellular Manufacture; *RE*—Robotics Equipment; *FMS*—Flexible Manufacturing Systems; *AA*—Automated Assembly; *CIM*—Computer-Integrated Manufacture. It presents the set of factors that may influence all these systems, from which, for each system, one subset must be highlighted. In this application, five sections are also considered to translate the conditions of each factor, and a new system performance indicator is defined, which was called **performance coefficient**.

Although the application has directed the attention towards the feasibility analysis of the implementation of a Flexible Manufacturing System (FMS), it presents elements which, complemented by the chapter called Justifying Capital Investment of [22], enables to study the feasibility of any of the other modern manufacturing systems.

Several exercises are proposed to analyze the feasibility of implementation of other modern manufacturing systems, besides the feasibility analysis of the FMS, which was studied in the text. There is an exercise to examine the feasibility of implementation of the CIM and another one to analyze the feasibility of implementation of the GT/CM. Besides that, another exercise proposes a comparative study of these three manufacturing systems, provided with advanced technology.

In application 6.5, the PDM is applied as a prediction of diagnoses [49], which is also decision making. In fact, diagnosing is nothing more than deciding, among the available options, which one of them is the most likely or the one with most evidence. This is another way to apply the PDM and which, therefore, shows one more diversification of the method. It also emphasizes that it may be implemented even in the analysis of problems that involve very large databases, without jeopardizing the promptness of the answer. We focused on the problem of the prediction of a medical diagnosis, although the method may be applied identically to the prediction of other diagnoses, such as defects in industrial machines, airplanes, ships, cars, trucks, etc.

There are several proposed exercises. All of them are based on the assembly of the CP of the PDM for a case less labor-intensive than the one analyzed in the text, as it was limited to only ten diseases and ten symptoms. With these several exercises, the intention is to enable the reader to assimilate the process well and to be able to verify its various possibilities.

6.1 Decision Concerning the Opening of a New Education Course by an Education Institution

As the first example, the application of the Paraconsistent Decision Method (PDM) will be studied in a problem the university managers constantly face: studying the feasibility of the opening of a new course in a given region. The factors that influence this decision are legal, social, economic and other factors. The most influential factors in these decisions will be chosen. As there is no need for much detailing in the results, only three sections for each one will be established. On the other hand, as the influences of the chosen factors on the feasibility of the course are practically the same, we will admit the all the factors have the same weight.

Then, by means of applications of the maximization (**MAX**) and minimization (**MIN**) techniques of logic E_τ (see Sect. 2.7), we reach a value of favorable evidence $(a_{i,R})$ and a value of contrary evidence $(b_{i,R})$, resultant, for each factor, which plotted in decision lattice τ, enable us to verify how each factor influences the decision (see Sect. 4.2.6). For the final decision making of the KE, knowing how each factor influences the enterprise is not sufficient, but it is interesting to know the joint influence of all the analyzed factors.

This is determined by the center of gravity (**W**) of the points that represent the factors, separately, in lattice τ. The degree of favorable evidence (a_W) of **W** is the arithmetic average of the favorable evidences resulting from all the factors $(a_{i,R})$, and its degree of contrary evidence (b_W) is the arithmetic average of the resultant contrary evidences for the factors $(b_{i,R})$. With these values, we are able to verify to which region of lattice τ the center of gravity **W** belongs, or to calculate the degree of certainty of W, apply the decision rule (see Sects. 4.2.7 and 4.2.8) make the decision.

6.1.1 Establishment of the Requirement Level

For this example, the established requirement level is 0.60. With that, it is established that the decision will only be favorable if, at the end, the degree of favorable evidence outweighs the degree of contrary evidence in at least 0.60. With this requirement level, the para-analyzing algorithm and the decision rule are the ones presented in Fig. 6.1.

H ≥ 0.60 ⇒ favorable decision (the opening of the course is feasible);

H ≤ − 0.60 ⇒   unfavorable decision (the opening of the course is infeasible);

−0.60 < H < 0.60 ⇒ inconclusive analysis.

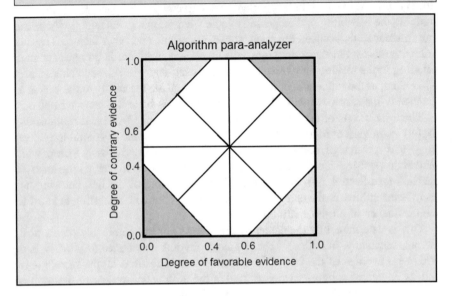

Fig. 6.1 Decision rule and para-analyzer algorithm for requirement level equal to 0.60

6.1.2 Selection of the Influence Factors and Establishment of the Sections

To facilitate the explanation, it will be considered during the text that we are studying the feasibility of the opening of a new course x (social communications course, for example) in a region Y (region of Ribeirão Preto, for example).

Several factors influence the success (or failure) of a new course. The twelve factors F_i (from F_1 to F_{12}) were chosen, deemed to be the ones with most influence on the feasibility of opening of a course in a certain region. For each one of these factors, three sections S_j (S_1 to S_3) were established, so that:

S_1 translates conditions in which the factor is **favorable** to the success of the course;

S_2 translates conditions in which the factor is **indifferent** to the success of the course; and

S_3 translates conditions in which the factor is **unfavorable** to the success of the course.

The chosen factors and the established sections are the following:

F_{01}: applicants/place ratio (C/V) of course X in the selection exams of region Y.

$$S_1 : C/V > 4; \quad S_2 : 2 \leq C/V \leq 4; \quad S_3 : C/V < 2.$$

F_{02}: number of high school graduating students (N_c) in region Y.

$$S_1 : N_c > 2V; \quad S_2 : V \leq N_c \leq 2V; \quad S_3 : N_c < V.$$

V = number of places offered for higher education in region Y.
F_{03}: number of jobs (N_e) offered annually in region Y.

$$S_1 : N_e > 2F; \quad S_2 : F \leq N_e \leq 2F; \quad S_3 : N_e < F.$$

F = annual number of graduates in higher education in region Y.
F_{04}: average monthly family income (R_f) of the population in region Y.

$$S_1 : R_f > R\$6,000.00; \quad S_2 : R\$2,000.00 \leq R_f \leq R\$6,000.00;$$
$$S_3 : R_f < R\$2,000.00.$$

F_{05}: annual average index (I_a) of course dropouts.

$$S_1 : I_a < 10\%; \quad S_2 : 10\% \leq I_a \leq 40\%; \quad S_3 : I_a > 40\%.$$

F_{06}: population density (DD) of the region.
S_1: high: DD > 400 inhabitants/km^2;
S_2: average: 100 inhabitants/km^2 $\leq$ DD $\leq$ 400 inhabitants/km^2;
S_3: low: DD < 100 inhabitants/km^2.
F_{07}: cost of the investments in fixed assets (C_{af}).

$$S_1 : C_{af} < 75\%R_a; \quad S_2 : 75\%R_a \leq C_{af} < 125\%R_a; \quad S_3 : C_{af} > 125\%R_a.$$

R_a = annual revenue foreseen for course X.
F_{08}: Concept of the institution with the community

$$S_1 : \text{Concept A or B}; \quad S_2 : \text{Concept C}; \quad S_3 : \text{Concept D or E}.$$

F_{09}: monthly cost with professors (C_{mp}).

$$S_1 : C_{mp} < 40\%R_m; \quad S_2 : 40\%R_m \leq C_{mp} \leq 70\%R_m; \quad S_3 : C_{mp} > 70\%R_m.$$

R_m = monthly revenue foreseen with course X.
F_{10}: value of the monthly fee (M_c) of course X.

$$S_1 : M_c < 80\%M_m; \quad S_2 : 80\%M_m \leq M_c \leq 120\%M_m; \quad S_3 : M_c > 120\%M_m.$$

M_m = average monthly fee of course X (or similar courses) in the other schools of region Y.

F_{11}: average number of students per class (N_{as}).

$$S_1 : N_{as} > 80; \quad S_2 : 50 \leq N_{as} \leq 80; \quad S_3 : N_{as} < 50.$$

F_{12}: average number of employees per class (N_{ft}).

$$S_1 : N_{ft} < 5; \quad S_2 : 5 \leq N_{ft} \leq 10; \quad S_3 : N_{ft} > 10.$$

6.1.3   Construction of the Database

To build the database, the KE chooses the experts and collects their opinions, through the values of favorable evidence (degrees of belief) (**a**) and contrary evidence (degrees of disbelief) (**b**) each one attributes to the success of the course, when each one of the conditions translated by the established sections are met for each one of the chosen factors. That is, for each section of each factor, which are the degrees of favorable evidence and contrary evidence each one of the experts attributes?

Besides that, the specialists are requested to attribute weights to the factors according to the importance of each one in the decision for the opening or not of course X in region Y. In this case, it is being admitted that all the factors have the same weight, equal to 1 (one).

In this application, it was supposed that the opinions of four specialists were collected (E_1: sociologist; E_2: economist; E_3: educator; E_4: business administrator) and that theur opinions are translated by Table 6.1.

It must be observed that this table constitutes a database, which may be used for the feasibility analysis of the opening of different courses, in different regions. In this example, it will be utilized to study the feasibility of course X in region Y.

6.1.4   Field Research

The following step is to conduct research in region Y about course X, to verify in which section each factor is, that is, which is the real condition of each factor. The result of this research may be summarized in Table 6.2.

With the results obtained in the research (Table 6.2), we can extract from the database (Table 6.1) the opinions of the specialists concerning the success of course

Table 6.1 Database: values of the favorable and contrary evidences attributed by the specialists to each one of the factors, in the conditions of the established sections (matrix of the annotations)

Data base			Group A				Group B			
			Spec 1		Spec 2		Spec 3		Spec 4	
Factor	Section	F and S	$a_{i,1}$	$b_{i,1}$	$a_{i,2}$	$b_{i,2}$	$a_{i,3}$	$b_{i,3}$	$a_{i,4}$	$b_{i,4}$
F_{01}	S_1	F01S1	1.0	0.0	0.9	0.1	1.0	0.2	0.8	0.3
	S_2	F01S2	0.9	0.2	0.6	0.3	0.8	0.4	0.5	0.6
	S_3	F01S3	0.7	0.4	0.3	0.8	0.6	0.5	0.2	0.9
F_{02}	S_1	F02S1	0.9	0.2	0.8	0.2	1.0	0.3	0.8	0.1
	S_2	F02S2	0.8	0.3	0.6	0.4	0.8	0.4	0.6	0.4
	S_3	F02S3	0.3	0.7	0.1	0.9	0.5	0.7	0.0	0.9
F_{03}	S_1	F03S1	0.9	0.1	0.8	0.2	1.0	0.1	0.7	0.3
	S_2	F03S2	0.6	0.4	0.6	0.5	0.7	0.3	0.5	0.5
	S_3	F03S3	0.3	0.6	0.5	0.8	0.4	0.7	0.0	0.9
F_{04}	S_1	F04S1	0.9	0.2	1.0	0.1	0.8	0.3	0.7	0.4
	S_2	F04S2	0.6	0.3	0.7	0.4	0.7	0.6	0.5	0.6
	S_3	F04S3	0.2	0.8	0.3	0.7	0.1	0.8	0.1	0.9
F_{05}	S_1	F05S1	0.9	0.1	0.8	0.2	1.0	0.1	0.8	0.3
	S_2	F05S2	0.7	0.2	0.6	0.5	0.7	0.3	0.6	0.5
	S_3	F05S3	0.1	0.8	0.5	0.8	0.4	0.7	0.0	0.9
F_{06}	S_1	F06S1	0.8	0.1	0.8	0.2	1.0	0.3	0.7	0.1
	S_2	F06S2	0.6	0.3	0.6	0.4	0.8	0.4	0.5	0.4
	S_3	F06S3	0.3	0.7	0.1	0.9	0.5	0.7	0.0	0.9
F_{07}	S_1	F07S1	1.0	0.0	1.0	0.1	0.8	0.3	0.8	0.4
	S_2	F07S2	0.8	0.3	0.7	0.4	0.7	0.6	0.5	0.6
	S_3	F07S3	0.5	0.8	0.3	0.7	0.1	0.8	0.2	0.9
F_{08}	S_1	F08S1	1.0	0.0	0.9	0.1	1.0	0.2	0.8	0.3
	S_2	F08S2	0.8	0.3	0.6	0.3	0.8	0.4	0.5	0.6
	S_3	F08S3	0.5	0.8	0.3	0.8	0.6	0.5	0.2	0.9
F_{09}	S_1	F09S1	1.0	0.0	0.9	0.2	1.0	0.1	0.8	0.3
	S_2	F09S2	0.8	0.3	0.7	0.4	0.8	0.2	0.5	0.6
	S_3	F09S3	0.3	0.9	0.2	0.9	0.5	0.6	0.0	0.9
F_{10}	S_1	F10S1	0.9	0.0	0.8	0.2	1.0	0.3	0.8	0.1
	S_2	F10S2	0.5	0.5	0.6	0.4	0.8	0.4	0.6	0.4
	S_3	F10S3	0.3	0.8	0.1	0.9	0.5	0.7	0.0	0.9
F_{11}	S_1	F11S1	1.0	0.0	1.0	0.1	1.0	0.0	0.8	0.2
	S_2	F11S2	0.8	0.3	0.7	0.4	0.9	0.2	0.5	0.4
	S_3	F11S3	0.2	0.9	0.3	0.3	0.5	0.5	0.0	1.0
F_{12}	S_1	F12S1	1.0	0.1	0.8	0.2	1.0	0.3	0.8	0.1
	S_2	F12S2	0.7	0.2	0.6	0.4	0.8	0.4	0.6	0.4
	S_3	F12S3	0.4	0.6	0.1	0.9	0.5	0.7	0.0	0.9

Table 6.2 Research results for course X in region Y (researched matrix)

Factor F$_i$	F01	F02	F03	F04	F05	F06	F07	F08	F09	F10	F11	F12
Section S$_i$	S1	S1	S3	S3	S2	S1	S3	S1	S2	S1	S3	S2

Table 6.3 Degrees of favorable evidence and contrary evidence to the success of course X, attributed by the experts, for the conditions of the influence factors in region Y (matrix of the researched data, M_{Dqp})

Calculation table									
1	2	3	4	5	6	7	8	9	10
Fi	Spi	Group A				Group B			
		Spec 1		Spec 2		Spec 3		Spec 4	
		$a_{i,1}$	$b_{i,1}$	$a_{i,2}$	$b_{i,2}$	$a_{i,3}$	$b_{i,3}$	$a_{i,4}$	$b_{i,4}$
F01	S1	1.0	0.0	0.9	0.1	1.0	0.2	0.8	0.3
F02	S1	0.9	0.2	0.8	0.2	1.0	0.3	0.8	0.1
F03	S3	0.3	0.6	0.5	0.8	0.4	0.7	0.1	0.9
F04	S3	0.2	0.8	0.3	0.7	0.1	0.8	0.0	0.9
F05	S2	0.7	0.2	0.6	0.5	0.7	0.3	0.5	0.5
F06	S1	0.8	0.1	0.8	0.2	1.0	0.3	0.8	0.1
F07	S3	0.5	0.8	0.3	0.7	0.1	0.8	0.0	0.9
F08	S1	1.0	0.0	0.9	0.1	1.0	0.2	0.8	0.3
F09	S2	0.8	0.3	0.7	0.4	0.8	0.2	0.5	0.6
F10	S1	0.9	0.0	0.8	0.2	1.0	0.3	0.8	0.1
F11	S3	0.2	0.9	0.3	0.9	0.5	0.5	0.0	1.0
F12	S2	0.7	0.2	0.6	0.4	0.8	0.4	0.6	0.4

X in the real conditions in which the factors are in region Y. This extraction is conducted by the CP of the PDM. Table 6.3 summarizes the values extracted from the database.

6.1.5   Obtention of the Resultant Degrees of Favorable and Contrary Evidences for the Factors

After the values of the database, we must apply the maximization (**MAX** operator) and minimization (**MIN** operator) rule of the paraconsistent annotated evidential logic to the experts' opinions, for each one of the chosen factors, in the sections obtained in the research. That is also performed by the CP of the PDM.

In the application of these rules, it is convenient that the groups are constituted, observing the specialists' majors (see Sects. 2.7 and 4.2.6). The previously exemplified experts and the following constitution of the groups will be utilized, considered the most adequate: Group A: sociologist (E_1) and economist (E_2), and Group B: educator (E_3) and business administrator (E_4).

This way, for the application of the maximization (**MAX**) and minimization (**MIN**) rules to these experts' opinions, we must do:

$$\mathbf{MIN}\{\mathbf{MAX}[(E_1), (E_2)], \mathbf{MAX}[(E_3), (E_4)]\} \text{ or } \mathbf{MIN}\{G_A, G_B\}$$

This way, we obtain, for each factor, in the researched section, the combined conclusion of the experts' opinions. They are the resultant degrees of favorable ($a_{i,R}$) and contrary ($b_{i,R}$) evidences for the factors.

As already said, the CP of the PDM performs all the operations: search for the values of favorable and contrary evidences in the database, once the result of the field research is known (which resulted in Table 6.4 or in columns from 3 to 10 of Tables 6.4 and 6.5); application of the maximization (**MAX** operator) (columns 11–14 of Tables 6.4 and 6.5) and minimization (**MIN** operator) (columns 15 and 16 of Tables 6.4 and 6.5) rules to the researched data in region Y in relation to course X (Table 6.3), obtaining the resultant values of the favorable and contrary evidences to the factors (columns 15 and 16 of Tables 6.4 and 6.5). Therefore, the degrees of favorable ($a_{i,R}$) and contrary ($b_{i,R}$) evidences, resultant to all the factors in the conditions translated by the researched sections, were obtained by the CP of the PDM and are presented in columns 15 and 16 of Tables 6.4 and 6.5.

From these values, the CP of the PDM calculates, for each factor F_i, the degree of certainty and the degree of uncertainty ($H_i = a_{i,R} - b_{i,R}$ and $G_i = a_{i,R} + b_{i,R} - 1$) (columns 17 and 18 of Tables 6.4 and 6.5).

6.1.6 Obtention of the Degrees of Favorable and Contrary Evidences of the Center of Gravity

The degrees of favorable (a_w) and contrary (b_w) evidences of the center of gravity are calculated by the weighted averages of the resultant degrees of favorable ($a_{i,R}$) evidence and contrary ($b_{i,R}$) evidence for the factors, adopting the weights that the specialists attributed to the factors (see Eq. 4.4). In this case, as it was admitted that all the factors have the same weight, equal to 1 (one), the referred weighted average is reduced to a simple arithmetic average (see Eq. 4.5) and the center of gravity coincides with the geometrical center of the points that represent the factors in the cartesian plane.

These values (a_w and b_w) are also calculated by the CP of the PDM and appear in the last line of columns 15 and 16 of Tables 6.4 and 6.5. From these values, the PDM calculates the degree of certainty of the center of gravity ($H_W = a_W - b_W$), which appears in the last line of column 17 of Tables 6.4 and 6.5.

The values of the degrees of certainty, which appear in column 17 of Tables 6.4 and 6.5, enable us to verify how each factor influences on the feasibility of the

Table 6.4 Degrees of favorable and contrary evidence to the conditions of the factors in Y region (researched data matrix, M_{Dpq})

Calculation table

1	2	3	4	5	6	7	8	9	10	11	12	13	14	15	16	17	18	19
Fi	Spi	Group A		Spec 2		Group B		Spec 4		A	E1 MAX E2	B	E3 MAX E4	A MIN B		Level of requirement		0.60
		Spec 1				Spec 3										Conclusions		
		ai,1	bi,1	ai,2	bi,2	ai,3	bi,3	ai,4	bi,4	ai,gA	bi,gA	ai,gB	bi,gB	ai,R	bi,R	H	G	Decision
F01	S1	1.0	0.0	0.9	0.1	1.0	0.2	0.8	0.3	1.0	0.0	1.0	0.3	1.0	0.3	0.7	0.3	VIABLE
F02	S1	0.9	0.2	0.8	0.2	1.0	0.3	0.8	0.1	0.9	0.2	1.0	0.1	0.9	0.2	0.7	0.1	VIABLE
F03	S3	0.3	0.6	0.5	0.8	0.4	0.7	0.1	0.9	0.5	0.6	0.4	0.4	0.4	0.6	-0.2	0.0	NOT CONCLUSIVE
F04	S3	0.2	0.8	0.3	0.7	0.1	0.8	0.0	0.9	0.3	0.7	0.1	0.1	0.1	0.7	-0.6	-0.2	UNVIABLE
F05	S2	0.7	0.2	0.6	0.5	0.7	0.3	0.5	0.5	0.7	0.2	0.7	0.5	0.7	0.5	0.2	0.2	NOT CONCLUSIVE
F06	S1	0.8	0.1	0.8	0.2	1.0	0.3	0.8	0.1	0.8	0.1	1.0	0.1	0.8	0.1	0.7	-0.1	VIABLE
F07	S3	0.5	0.8	0.3	0.7	0.1	0.8	0.0	0.9	0.5	0.7	0.1	0.1	0.1	0.7	-0.6	-0.2	UNVIABLE
F08	S1	1.0	0.0	0.9	0.1	1.0	0.2	0.8	0.3	1.0	0.0	1.0	0.3	1.0	0.3	0.7	0.3	VIABLE
F09	S2	0.8	0.3	0.7	0.4	0.8	0.2	0.5	0.6	0.8	0.3	0.8	0.6	0.8	0.6	0.2	0.4	NOT CONCLUSIVE
F10	S1	0.9	0.0	0.8	0.2	1.0	0.3	0.8	0.1	0.9	0.0	1.0	0.1	0.9	0.1	0.8	0.0	VIABLE
F11	S3	0.2	0.9	0.3	0.9	0.5	0.5	0.0	1.0	0.3	0.9	0.5	0.5	0.3	0.9	-0.6	0.2	UNVIABLE
F12	S2	0.7	0.2	0.6	0.4	0.8	0.4	0.6	0.4	0.7	0.2	0.8	0.4	0.7	0.4	0.3	0.1	NOT CONCLUSIVE
Baricenter W: averages of the resultant degrees														0.64	0.45	0.19	0.09	NOT CONCLUSIVE

course and the last one, of the center of gravity, translates the joint impact of the factors and enables the final decision making about the opening of course X in region Y.

6.1.7 Analysis of the Results

The final results, after the application of the maximization and minimization rules, will be analyzed, firstly, by the application of the decision rule and, then, by the para-analyzing algorithm (see Sect. 6.1.1).

To apply the decision rule, you just need to have the degree of certainty and compare it to the requirement level. This is performed by the CP of the PDM in column 19 of Table 6.4. Therefore, observing the results of this column, it is verified that five factors, F_{01}, F_{02}, F_{06}, F_{08} and F_{10}, indicate that the enterprise is feasible, that is, recommend the opening of course X in region Y, as their degrees of certainty resulted higher or equal to 0.60, which was the established requirement level; three factors, F_{04}, F_{07} and F_{11}, indicate that the enterprise is infeasible, that is, recommend the non-opening of course X in region Y, as their degrees of certainty resulted lower or equal to -0.60; and finally, four factors, F_{03}, F_{05}, F_{09} and F_{12}, are inconclusive, as their degrees of certainty resulted between -0.60 and 0.60. These last ones do not recommend nor avoid the recommendation to the opening of course X in region Y.

The influence of the twelve factors together is translated by the center of gravity. As their degree of certainty resulted equal to 0.19 (last line of column 17 of Table 6.4), it is inferred that the conducted analysis presented inconclusive result, as $-0.60 < 0.19 < 0.60$. Therefore, the analysis, being inconclusive, only suggests that, if there is interest or any doubt, new studies are conducted so that the doubt is clarified.

The decision by the para-analyzing algorithm is performed by plotting the degrees of favorable evidence and contrary evidence resulting from the decision lattice τ, and verifying to which regions the representative points of the factors and the center of gravity belong. That is also performed by the CP of the PDM, and is shown in Fig. 6.2a.

In Fig. 6.2a, five factors, F_{01}, F_{02}, F_{06}, F_{08} and F_{10}, are in the truth region, suggesting favorable decision, that is, feasibility of the opening of the course (in Fig. 6.2a, only four points appear, because $F_{01} \equiv F_{08} = (1.0; 0.3)$, once their coordinates (degrees of favorable and contrary evidence, resultant) are equal); three factors, F_{04}, F_{07} and F_{11}, are in the falsity region, suggesting unfavorable decision, that is, unfeasibility of the opening of the course [in the figure only two points appear, because $F_{04} \equiv F_{07} = (0.1; 0.7)$]; and four factors, F_{03}, F_{05}, F_{09} and F_{12}, in a distinct region from the previous ones, being inconclusive. It must be observed that factors F_{04}, F_{07} and F_{11} belong to the falsity limit line (which, by convention, belongs to the falsity region), as they have a degree of certainty (-0.60), in the module, equal to the adopted requirement level.

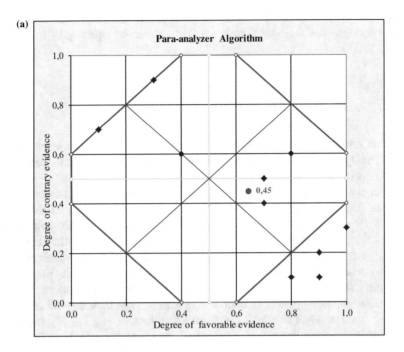

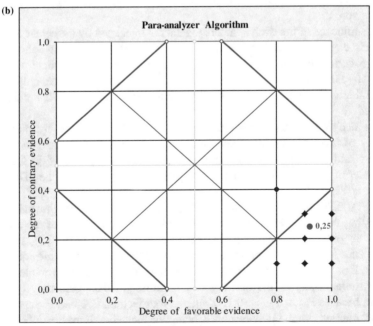

Fig. 6.2 a Application of the para-analyzing algorithm for feasibility analysis of course X in region Y. **b** Application of the para-analyzing algorithm for feasibility analysis of course X in region Y, when all the factors are favorable (section S_1), to the requirement level 0.60

Table 6.5 Calculation table of the CP of PDM in the analysis of viability of the X course in the Y region, when all factors are in condiction S_1

Calculation table

1	2	3	4	5	6	7	8	9	10	11	12	13	14	15	16	17	18	19
Fi	Spi	Group A				Group B				A		B		A MIN B		Level of requirement		0.60
										E1 MAX E2		E3 MAX E4				Conclusions		
		Spec 1		Spec 2		Spec 3		Spec 4										
		$ai,1$	$bi,1$	$ai,2$	$bi,2$	$ai,3$	$bi,3$	$ai,4$	$bi,4$	ai,gA	bi,gA	ai,gB	bi,gB	ai,R	bi,R	H	G	Decision
F01	S1	1.0	0.0	0.9	0.1	1.0	0.2	0.8	0.3	1.0	0.0	1.0	0.3	1.0	0.3	0.7	0.3	VIABLE
F02	S1	0.9	0.2	0.8	0.2	1.0	0.3	0.8	0.1	0.9	0.2	1.0	0.1	0.9	0.2	0.7	0.1	VIABLE
F03	S1	0.9	0.1	0.8	0.2	1.0	0.1	0.7	0.3	0.9	0.1	1.0	0.3	0.9	0.3	0.6	0.2	VIABLE
F04	S1	0.9	0.2	1.0	0.1	0.8	0.3	0.7	0.4	1.0	0.1	0.8	0.4	0.8	0.4	0.4	0.2	NOT CONCLUSIVE
F05	S1	0.9	0.1	0.8	0.2	1.0	0.1	0.7	0.3	0.9	0.1	1.0	0.3	0.9	0.3	0.6	0.2	VIABLE
F06	S1	0.8	0.1	0.8	0.2	1.0	0.3	0.8	0.1	0.8	0.1	1.0	0.1	0.8	0.1	0.7	-0.1	VIABLE
F07	S1	1.0	0.0	1.0	0.1	0.8	0.3	0.7	0.4	1.0	0.0	0.8	0.4	0.8	0.4	0.4	0.2	NOT CONCLUSIVE
F08	S1	1.0	0.0	0.9	0.1	1.0	0.2	0.8	0.3	1.0	0.0	1.0	0.3	1.0	0.3	0.7	0.3	VIABLE
F09	S1	1.0	0.0	0.9	0.2	1.0	0.1	0.8	0.3	1.0	0.0	1.0	0.3	1.0	0.3	0.7	0.3	VIABLE
F10	S1	0.9	0.0	0.8	0.2	1.0	0.3	0.8	0.1	0.9	0.0	1.0	0.1	0.9	0.1	0.8	0.0	VIABLE
F11	S1	1.0	0.0	1.0	0.1	1.0	0.0	0.8	0.2	1.0	0.0	1.0	0.2	1.0	0.2	0.8	0.2	VIABLE
F12	S1	1.0	0.1	0.8	0.2	1.0	0.3	0.8	0.1	1.0	0.1	1.0	0.1	1.0	0.1	0.9	0.1	VIABLE
Baricenter W: averages of the resultant degrees														0.92	0.25	0.67	0.17	VIABLE

The disparate influences of all these factors on the decision of the feasibility of course X in region Y may be summarized by the center of gravity W of the points that represent them. As W is in the region of quasi-truth, tending to inconsistency, it is deduced that the final result of the analysis is inconclusive. That is, the analysis does not recommend the opening of course X in region Y, but does not exclude this possibility either. It only suggests that, if it is of interest, new researches are conducted, in an attempt to increase the evidences.

6.1.8   Analysis of the Feasibility of Course X in Region Y, in Another Scenario

To carry out a test of the reliability of the PDM and an exercise of its application, the viability of a course X in region Y was analyzed, assuming that in the field research, all the factors were in the conditions of Section S_1, that is, all the factors were favorable to course X in region Y. In this case, evidently, it was not a surprise that the analysis led to the conclusion for the feasibility of course X in region Y.

In fact, applying the CP of the PDM to this case, that is, placing S_1 in all the lines of column 2 of Table 6.4 of the PDM, we obtain $a_W = 0.92$ and $b_W = 0.25$ (Table 6.5). That enables us to calculate $H_W = a_W - b_W = 0.92 - 0.25 = 0.67$.

As $0.67 \geq 0.60$, the decision rule enables us to infer for the feasibility of course X in region Y to the requirement level 0.60, in this new scenario. The para-analyzing algorithm starts having the aspect represented in Fig. 6.2b.

Observing the results of columns 15 and 16 of Table 6.5, we notice that in Fig. 6.2b coincidences of points representative of factors occur. They are:

$$F_{01} \equiv F_{08} \equiv F_{09} = (1.0; 0.3); \quad F_{03} \equiv F_{05} = (0.9; 0.3) \text{ and } F_{04} \equiv F_{07} = (0.8; 0.4).$$

The other points are isolated. The center of gravity is

$$W = (0.92; 0.25).$$

6.2   Feasibility Analysis of the Launch of a Product

In this case, the PDM will be applied in a problem the marketing professionals constantly face: studying the viability of the launch of a new product. In sequence, it will be seen how the stages of the PDM are, since the establishment of the requirement level until the final decision, which may be made based on the decision rule or on the para-analyzer algorithm.

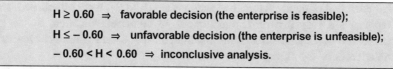

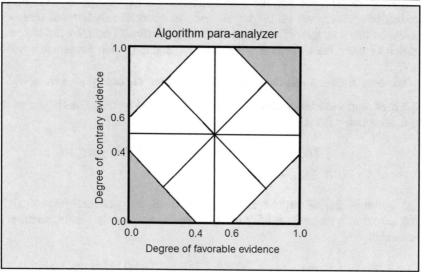

Fig. 6.3 Decision rule and para-analyzer algorithm for requirement level 0.60

6.2.1   Establishment of the Requirement Level

For this application example, the established requirement level is 0.60. It is a medium requirement level, but it is justified, taking into account the fact of being a product in which the decision of launching it does not demand too much investment, does not represent major economic, social or human responsibilities. That is, the favorable decision to the product launch does not put high amounts, human lives, the environment, etc., at risk.

With the establishment of the requirement level, the decision rule and the para-analyzer algorithm are automatically configured (Fig. 6.3).

6.2.2   Selection of the Influence Factors and Establishment of the Sections

The ten factors F_i; (F_{01} to F_{10}) that have the greatest influence on the feasibility of the launch of a product were chosen.

For each one of these factors, five sections S_j (S_1 to S_5) were established, so that S_1 represents a **very favorable** situation; S_2, a **favorable** situation; S_3, an **indifferent** situation; S_4, an **unfavorable** situation; and S_5, a **very unfavorable** situation to the success of the enterprise (product launch).

It must be observed that the characterization of the sections depends on the product being launched, on market analysis, on economic studies and other elements. In this example, this characterization is performed without the strict utilization of these items, as it is a theoretical example for the presentation of the method.

The chosen factors and the established ranges are the ones presented below:

F_{01}: need and usefulness of the product—Translated by the percentage π of the population that utilizes it.

$$S_1 : \pi > 90\%; \quad S_2 : 70\% < \pi \leq 90\%; \quad S_3 : 30\% \leq \pi \leq 70\%;$$
$$S_4 : 10\% \leq \pi \leq 30\%; \quad S_5 : \pi < 10\%.$$

F_{02}: quantity (η) of attributes or functions of the product—Measured by comparison with the average M of attributes or functions of the similar products in the market.

$$S_1 : \eta > 1.5M; \quad S_2 : 1.2M < \eta \leq 1.5M; \quad S_3 : 0.8M \leq \eta \leq 1.2M;$$
$$S_4 : 0.5M \leq \eta < 0.8M; \quad S_5 : \eta < 0.5M.$$

F_{03}: competition—Translated by the quality and quantity of competitors in the region

$$S_1 : \text{very small}; \quad S_2 : \text{small}; \quad S_3 : \text{medium};$$
$$S_4 : \text{large}; \quad S_5 : \text{very large}.$$

F_{04}: customers' potential—Translated by the size and purchase power of the region population.

$$S_1 : \text{very large}; \quad S_2 : \text{large}; \quad S_3 : \text{medium};$$
$$S_4 : \text{small}; \quad S_5 : \text{very small}.$$

F_{05}: acceptance of the product or similar product already existent in the market—Translated by the percentage π of the population that utilizes it.

$$S_1 : \pi > 90\%; \quad S_2 : 70\% < \pi \leq 90\%; \quad S_3 : 30\% \leq \pi \leq 70\%;$$
$$S_4 : 10\% \leq \pi \leq 30\%; \quad S_5 : \pi < 10\%.$$

F$_{06}$: product price (φ) in the market—Translated according to the average price P of an equal product (or similar products) already existent in the market.

$S_1 : \phi < 70\%P;\quad S_2 : 70\%P \leq \phi < 90\%P;\quad S_3 : 90\%P \leq \phi \leq 110\%P;$
$S_4 : 110\%P < \phi \leq 130\%P;\quad S_5 : \phi > 130\%P.$

F$_{07}$: estimated product cost (θ)—Translated according to the average price P of a same product (or similar products) already existent in the market.

$S_1 : \theta < 20\%P;\quad S_2 : 20\%P \leq \theta < 40\%P;\quad S_3 : 40\%P \leq \theta \leq 60\%P;$
$S_4 : 60\%P < \theta \leq 80\%P;\quad S_5 : \theta > 80\%P.$

F$_{08}$: product life cycle (C)—Measured in a time unit T (one year, for example).

$S_1 : C > 10T;\quad S_2 : 8T < C \leq 10T;\quad S_3 : 4T \leq C \leq 8T;$
$S_4 : 2T \leq C < 4T;\quad S_5 : C < 2T.$

F$_{09}$: deadline (λ) for project development and product implementation—Measured according to the life cycle (C).

$S_1 : \lambda < 10\%C;\quad S_2 : 10\%C \leq \lambda < 30\%C;\quad S_3 : 30\%C \leq \lambda \leq 70\%C;$
$S_4 : 70\%C < \lambda \leq 90\%C;\quad S_5 : \lambda > 90\%C.$

F$_{10}$: investment (I) for project development and product implementation—Measured according to the expected net result (R) in the product life cycle.

$S_1 : I < 20\%R;\quad S_2 : 20\%R \leq I < 40\%R;\quad S_3 : 40\%R \leq I \leq 60\%R;$
$S_4 : 60\%R < I \leq 80\%R;\quad S_5 : I > 80\%R.$

6.2.3 Construction of the Database

We admit that four experts were selected: E_1—marketing professional; E_2—economist; E_3—production engineer; E_4—business administrator. We also admit that the weights attributed to the factors by the specialists are equal (all the factors have a weight equal to 1 (one), for example). Thus, the database (Table 6.6) will be limited to the matrix of the annotations, that is, to the degrees of favorable evidence and contrary evidence attributed by the specialists to the factors, in the conditions defined by the five sections.

Table 6.6 Database: degrees of the favorable evidence and contrary evidence attributed by the specialists to the factors, in the conditions defined by the sections

F_i	S_j	E_1		E_2		E_3		E_4	
		$a_{i,j,1}$	$b_{i,j,1}$	$a_{i,j,2}$	$b_{i,j,2}$	$a_{i,j,3}$	$b_{i,j,3}$	$a_{i,j,4}$	$b_{i,j,4}$
F_{01}	S_1	0.88	0.04	0.94	0.14	0.84	0.08	0.78	0.03
	S2	0.63	0.19	0.79	0.23	0.73	0.14	0.59	0.24
	S3	0.48	0.43	0.53	0.44	0.58	0.39	0.48	0.41
	S4	0.23	0.77	0.41	0.61	0.33	0.73	0.29	0.53
	S5	0.01	0.94	0.13	0.88	0.14	1.00	0.17	0.91
F_{02}	S_1	1.00	0.05	0.95	0.15	1.00	0.1	0.85	0.00
	S2	0.75	0.25	0.85	0.25	0.85	0.3	0.73	0.35
	S3	0.55	0.45	0.55	0.45	0.65	0.4	0.45	0.55
	S4	0.35	0.65	0.31	0.79	0.29	0.7	0.24	0.83
	S5	0.00	0.95	0.15	0.75	0.15	0.85	0.25	1.00
F_{03}	S_1	0.92	0.08	0.98	0.18	0.88	0.12	0.82	0.07
	S2	0.67	0.23	0.83	0.27	0.77	0.18	0.63	0.28
	S3	0.52	0.47	0.57	0.48	0.62	0.43	0.52	0.45
	S4	0.17	0.73	0.24	0.65	0.37	0.67	0.33	0.64
	S5	0.05	0.98	0.17	0.83	0.18	0.02	0.21	0.95
F_{04}	S_1	0.95	0.11	1.00	0.21	0.91	0.15	0.85	0.10
	S2	0.70	0.26	0.86	0.30	0.80	0.21	0.66	0.31
	S3	0.55	0.50	0.60	0.51	0.65	0.46	0.55	0.48
	S4	0.30	0.76	0.48	0.68	0.22	0.70	0.28	0.6
	S5	0.08	1.00	0.20	0.86	0.21	0.05	0.24	0.98
F_{05}	S_1	1.00	0.88	0.06	0.10	0.95	0.85	0.04	0.00
	S2	0.70	0.20	0.80	0.30	0.80	0.20	0.70	0.30
	S3	0.50	0.50	0.60	0.50	0.60	0.40	0.50	0.40
	S4	0.30	0.70	0.33	0.69	0.30	0.70	0.26	0.73
	S5	0.00	1.00	0.10	0.80	0.90	0.08	1.00	0.15
F_{06}	S_1	0.90	0.10	1.00	0.10	0.90	0.00	1.00	0.00
	S2	0.80	0.30	0.80	0.20	0.70	0.30	0.70	0.20
	S3	0.60	0.50	0.60	0.40	0.50	0.40	0.50	0.50
	S4	0.40	0.60	0.40	0.70	0.30	0.60	0.30	0.70
	S5	0.10	0.80	0.20	0.90	0.13	1.00	0.00	1.00
F_{07}	S_1	0.95	0.15	1.00	0.10	0.85	0.00	1.00	0.05
	S2	0.85	0.25	0.85	0.3	0.73	0.35	0.75	0.25
	S3	0.55	0.45	0.65	0.4	0.45	0.55	0.55	0.45
	S4	0.40	0.65	0.35	0.75	0.24	0.78	0.35	0.65
	S5	0.05	0.88	0.15	0.85	0.12	1.00	0.00	0.95
F_{08}	S_1	0.98	0.18	0.88	0.12	0.82	0.07	0.92	0.08
	S2	0.83	0.27	0.77	0.18	0.63	0.28	0.67	0.23
	S3	0.57	0.48	0.62	0.43	0.52	0.45	0.52	0.47

(continued)

Table 6.6 (continued)

F_i	S_j	E_1		E_2		E_3		E_4	
		$a_{i,j,1}$	$b_{i,j,1}$	$a_{i,j,2}$	$b_{i,j,2}$	$a_{i,j,3}$	$b_{i,j,3}$	$a_{i,j,4}$	$b_{i,j,4}$
		0.45	0.65	0.37	0.85	0.33	0.57	0.27	0.86
	S5	0.08	0.83	0.18	0.95	0.21	0.95	0.05	0.98
F_{09}	S_1	1.00	0.21	0.91	0.15	0.85	0.10	0.95	0.11
	S2	0.86	0.30	0.80	0.21	0.66	0.31	0.7	0.26
	S3	0.60	0.51	0.65	0.46	0.55	0.48	0.55	0.50
	S4	0.39	0.76	0.30	0.70	0.36	0.60	0.30	0.76
	S5	0.10	0.86	0.15	0.93	0.24	0.98	0.08	1.00
F_{10}	S_1	0.94	0.14	0.84	0.08	0.78	0.03	0.88	0.04
	S2	0.79	0.23	0.73	0.14	0.59	0.24	0.63	0.19
	S3	0.53	0.44	0.58	0.39	0.48	0.41	0.48	0.43
	S4	0.41	0.69	0.33	0.63	0.29	0.53	0.23	0.69
	S5	0.13	0.79	0.14	0.90	0.17	0.91	0.01	0.94

6.2.4   Field Research and Calculation of the Resultant Degrees of Favorable Evidence and Contrary Evidence for the Factors and Center of Gravity

We must conduct research about the product in the region where it will be launched, to verify in which section each factor is. The researchers must verify, in the region where the product will be launched, in which section S_j (with $1 \leq j \leq 5$) each one of the factors F_i (with $1 \leq i \leq 10$) that influence the feasibility of the product is. With the research results, S_{pi}, column 2 of Table 6.7 is filled out. After that, the CP of the PDM extracts from the database (Table 6.6) the experts' opinions concerning the feasibility of the product launch in the chosen region, in the conditions translated by the researched sections. These views are summarized in columns 3–10 of Table 6.7.

For the application of the operators, the groups must be constituted, observing the expert's majors. Within the selected specialists, a possible and adequate formation is group A—the marketing professional (E_1) with the economist (E_2); group B—the production engineer (E_3) with the business administrator (E_4). This way, for the application of the maximization (**MAX** operator) and minimization (**MIN** operator) techniques to these experts' opinions, we must do:

$$\mathbf{MIN\{MAX[E_1, E_2]; MAX[E_3, E_4]\}} \text{ or } \mathbf{MIN\{G_A; G_B\}}.$$

In Table 6.7, the results of the application of the **MAX** operator in groups A and B (intragroup) are in columns 11–14. The results of the application of the **MIN** operator among groups A and B (intergroup), which are the resultant degrees of

Table 6.7 Calculation table of the PDM

1	2	3	4	5	6	7	8	9	10	11	12	13	14	15	16	17	18	19
Fi	S_j	Group A				Group B				A		B		A MIN B		Conclusions		Level of requirement = 0.60
		E1		E2		E3		E4		E_1 MAX E_2		E_3 MAX E_4						
		$a_{i,1}$	$b_{i,1}$	$a_{i,2}$	$b_{i,2}$	$a_{i,3}$	$b_{i,3}$	$a_{i,4}$	$b_{i,4}$	$a_{i,gA}$	$b_{i,gA}$	$a_{i,gB}$	$b_{i,gB}$	$a_{i,R}$	$b_{i,R}$	H	G	Decision
F_{01}	S_5	0.01	0.94	0.13	0.88	0.14	1.00	0.17	0.91	0.13	0.88	0.17	0.91	0.13	0.91	−0.78	0.04	UNVIABLE
F_{02}	S_1	1.00	0.05	0.95	0.15	1.00	0.10	0.85	0.00	1.00	0.05	1.00	0.00	1.00	0.05	0.95	0.05	VIABLE
F_{03}	S_1	0.92	0.08	0.98	0.18	0.88	0.12	0.82	0.07	0.98	0.08	0.88	0.07	0.88	0.08	0.80	−0.04	VIABLE
F_{04}	S_2	0.70	0.26	0.86	0.30	0.80	0.21	0.66	0.31	0.86	0.26	0.80	0.21	0.80	0.26	0.54	0.06	NOT CONCLUSIVE
F_{05}	S_1	1.00	0.88	0.06	0.10	0.95	0.85	0.04	0.00	1.00	0.10	0.95	0.00	0.95	0.10	0.85	0.05	VIABLE
F_{06}	S_5	0.10	0.80	0.20	0.90	0.13	1.00	0.00	1.00	0.20	0.80	0.13	1.00	0.13	1.00	−0.87	0.13	UNVIABLE
F_{07}	S_4	0.40	0.65	0.35	0.75	0.24	0.78	0.35	0.65	0.40	0.65	0.35	0.65	0.35	0.65	0.30	0.00	NOT CONCLUSIVE
F_{08}	$S4$	0.45	0.65	0.37	0.85	0.33	0.57	0.27	0.86	0.45	0.65	0.33	0.57	0.33	0.65	−0.32	−0.02	NOT CONCLUSIVE
F_{09}	S_1	1.00	0.21	0.91	0.15	0.85	0.10	0.95	0.11	1.00	0.15	0.95	0.10	0.95	0.15	0.80	0.10	VIABLE
F_{10}	S_2	0.79	0.23	0.73	0.14	0.59	0.24	0.63	0.19	0.79	0.14	0.63	0.19	0.63	0.19	0.44	−0.18	NOT CONCLUSIVE
Baricenter W: averages of the resultant degrees														0.615	0.404	0.211	0.019	

favorable evidence and contrary evidence, appear in columns 15 and 16. This way, we obtain, for each factor, in the condition of the section obtained in the research, the combined conclusion of the experts' opinions. The degrees of certainty and uncertainty for each factor, in the condition of the researched section, appear in columns 17 and 18.

6.2.5   Analysis of the Results

Firstly, the analysis of the final results by the application of the decision rule will be conducted (see Sect. 6.2.1). The CP of the PDM had already done that when it compared the degrees of certainty of column 17 of Table 6.7 to the requirement level (0.60) and gave the result in column 19. Thus, we may observe that, in the researched conditions and to the requirement level 0.60, four factors, F_{02}, F_{03}, F_{05} and F_{09}, indicate that the product launch is feasible; two, F_{01} and F_{06}, indicate that the product launch in infeasible; and the other four, F_{04}, F_{07}, F_{08} and F_{10}, are inconclusive, that is, do not provide indication in favor nor against launching the product.

However, what matters is the joint influence of all the factors on the feasibility of the product launch, which is translated by the center of gravity W of the points that represent them, isolatedly. In the last line of columns 15 and 16 of Table 6.6, are the degrees of favorable evidence (a_W) and of contrary evidence (b_W) of the center of gravity, which enable the CP of the PDM to calculate the corresponding degree of certainty (last line of column 17) as follows: $H_W = a_W - b_W = 0.615 - 0.404 = 0.211$.

Considering that $-0.60 < 0.211 < 0.60$, applying the decision rule, the CP of the PDM itself already concludes that the analysis is inconclusive, that is, the analysis does not enable us to conclude for the feasibility nor the unfeasibility of the product launch in the chosen region.

This same analysis may be performed by the para-analyzer algorithm. For this purpose, we plot the resultant degrees of favorable evidence and contrary evidence, at lattice τ (Fig. 6.4), adopting as truth and falsity limit lines the lines determined by $|H| = |a - b| = 0.60$, and as inconsistency and paracompleteness limit lines, the lines determined by $|G| = |a + b - 1| = 0.60$, once the adopted requirement level was 0.60.

In the studied case, feasibility analysis of the product launch in the chosen region, the observation of the points that represent the influence factors in lattice τ shows that: four factors (F_{02}, F_{03}, F_{05}, and F_{09}) belong to the truth region (of favorable decision or feasibility), thus recommending the launch of the product to the requirement level 0.60; two factors (F_{01} and F_{06}) belong to the falsity region (unfeasibility), recommending the non-launch of the product.

The other factors belong to the low definition regions, indicating that the product launch is not feasible, but it is not unfeasible either. F_{04} belongs to the region of quasi-truth, tending to inconsistency; F_{10} belongs to the region of quasi-truth,

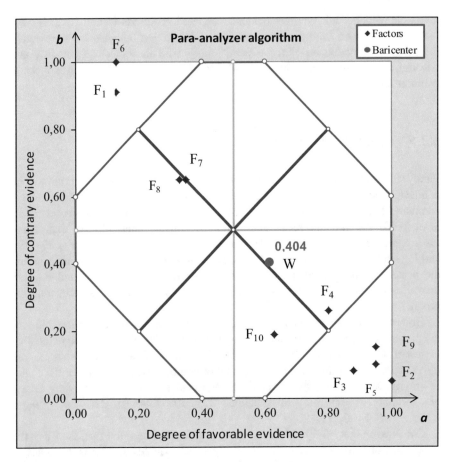

Fig. 6.4 Analysis of the result by the para-analyzer algorithm; in this figure, the factors, the center of gravity, the limit lines and the borders are represented

tending to paracompleteness; F_{08} belongs to the region of quasi-falsity, tending to paracompleteness; F_{07} belongs to the regions of quasi-falsity, tending to inconsistency and quasi-falsity, tending to paracompleteness. These last four factors are, then, inconclusive.

But the disparate influences of all these factors on the feasibility product launch in the chosen region may be summarized by the center of gravity **W**. This is the center of gravity of the points that represent the factors, isolatedly, and translates the joint influence of the ten analyzed factors. As **W** is in the region of quasi-truth, tending to inconsistency, it is deduced that the global result is: inconclusive analysis. That is, the analysis does not recommend the product launch, but does not exclude this possibility either. It only suggests that, if it is of interest, new researches are conducted, in an attempt to increase the evidences.

Once more, it is important to observe that, after the research is conducted, that is, column 2 of Table 6.7 is filled out, all the other operations, translated by columns 3–19 (search for the values in the database; application of the maximization and minimization rules, to obtain the resultant degrees of favorable and contrary evidences for the factors; calculation of the degrees of favorable and contrary evidences of the center of gravity; computation of the degrees of certainty and contradiction; application of the decision rule; and application of the para-analyzer algorithm), are automatically performed by the CP of the PDM.

To test the reliability of the PDM and exercise its application, the feasibility analysis of the launch of another product in another region will be conducted, admitting that in the field research, all the factors are in section S_1, that is, all the factors are very favorable to the launch. In this case, evidently, it is not a surprise that the analysis leads us to conclude for the feasibility of the product, which will certify the reliability of the method.

In fact, applying the PDM to this situation, we obtain $a_W = 0.933$ and $b_W = 0.093$. These values enable us to calculate $H_W = a_W - b_W = 0.933 - 0.093 = 0.840$. As $0.840 \geq 0.60$, the decision rule enables us to infer for the **feasibility** of the launch of this other product in the considered region (Table 6.8; Fig. 6.5).

On the contrary, if all the factors were very unfavorable (section S_5), the application of the PDM would lead to $a_W = 0.147$ and $b_W = 0.904$, enabling us to calculate the degree of certainty of the center of gravity. $H_W = a_W - b_W = 0.147 - 0.904 = -0.757$. As $-0.757 \leq -0.60$, the decision rule would lead us to infer for the **unfeasibility** of the launch of the product in the region. (Table 6.9; Fig. 6.6).

To verify the influence of the requirement level in the decision (which represents one more possibility of the PDM), a case was analyzed, in which five factors (F_{02}, F_{04}, F_{06}, F_{08} and F_{09}) are very favorable (section S_1) and other five (F_{01}, F_{03}, F_{05}, F_{07} and F_{10}) are only favorable (section S_2). Applying the PDM, it was obtained: $H_W = a_W - b_W = 0.846 - 0.159 = 0.687$ (Table 6.10). Then, if the requirement level is 0.60, the decision is favorable (the product is **feasible**), once $0.687 > 0.60$ (Fig. 6.7a); however, if the requirement level is 0.75, the decision is that the analysis is inconclusive, once ($-0.75 < 0.687 < 0.75$) (Table 6.11; Fig. 6.7b).

It must be observed that, in the passage from Tables 6.10 and 6.11, the only alteration occurred in column 19, as the decision depends on the requirement level. Analogously, when moving from Fig. 6.7 to Table 6.8, the alteration occurred only in the positions of the limit lines, which depend on the requirement level.

Table 6.8 Application of the PDM when all factors are in condition very favorable (section S_1)

1	2	3	4	5	6	7	8	9	10	11	12	13	14	15	16	17	18	19
F_i	S_j	Group A				Group B				A		B		A MIN B		Conclusions		Level of requirement = 0.60
		E_1		E_2		E_3		E_4		E_1 MAX E_2		E_3 MAX E_4						
		$a_{i,1}$	$b_{i,1}$	$a_{i,2}$	$b_{i,2}$	$a_{i,3}$	$b_{i,3}$	$a_{i,4}$	$b_{i,4}$	$a_{i,gA}$	$b_{i,gA}$	$a_{i,gB}$	$b_{i,gB}$	$a_{i,R}$	$b_{i,R}$	H	G	Decision
F_{01}	S_1	0.88	0.04	0.94	0.14	0.84	0.08	0.78	0.03	0.94	0.04	0.84	0.03	0.84	0.04	0.8	−0.12	VIABLE
F_{02}	S_1	1.00	0.05	0.95	0.15	1.00	0.10	0.85	0.00	1.00	0.05	1.00	0.00	1.00	0.05	0.95	0.05	VIABLE
F_{03}	S_1	0.92	0.08	0.98	0.18	0.88	0.12	0.82	0.07	0.98	0.08	0.88	0.07	0.88	0.08	0.80	−0.04	VIABLE
F_{04}	S_1	0.95	0.11	1.00	0.21	0.91	0.15	0.85	0.10	1.00	0.11	0.91	0.10	0.91	0.11	0.80	0.02	VIABLE
F_{05}	S_1	1.00	0.88	0.06	0.10	0.95	0.85	0.04	0.00	1.00	0.10	0.95	0.00	0.95	0.10	0.85	0.05	VIABLE
F_{06}	S_1	0.90	0.10	1.00	0.10	0.90	0.00	1.00	0.00	1.00	0.10	1.00	0.00	1.00	0.10	0.90	0.10	VIABLE
F_{07}	S_1	0.95	0.15	1.00	0.10	0.85	0.00	1.00	0.05	1.00	0.10	1.00	0.00	1.00	0.10	0.90	0.10	VIABLE
F_{08}	S_1	0.98	0.18	0.88	0.12	0.82	0.07	0.92	0.08	0.98	0.12	0.92	0.07	0.92	0.12	0.80	0.04	VIABLE
F_{09}	S_1	1.00	0.21	0.91	0.15	0.85	0.10	0.95	0.11	1.00	0.15	0.95	0.10	0.95	0.15	0.80	0.10	VIABLE
F_{10}	S_1	0.94	0.14	0.84	0.08	0.78	0.03	0.88	0.04	0.94	0.08	0.88	0.03	0.88	0.08	0.80	−0.04	VIABLE
Baricenter W: averages of the resultant degrees														0.933	0.093	0.840	0.026	VIABLE

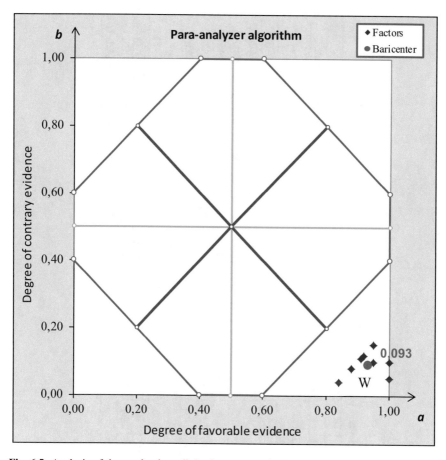

Fig. 6.5 Analysis of the result when all the factors are very favorable (section S_1)

Table 6.9 Application of the PDM when all factors are in condition very unfavorable (section S_5)

1	2	3	4	5	6	7	8	9	10	11	12	13	14	15	16	17	18	19
		Group A				Group B				A		B		A MIN B		Conclusions		Level of requirement = 0.60
F_i	S_j	E_1		E_2		E_3		E_4		E_1 MAX E_2		E_3 MAX E_4						
		$a_{i,1}$	$b_{i,1}$	$a_{i,2}$	$b_{i,2}$	$a_{i,3}$	$b_{i,3}$	$a_{i,4}$	$b_{i,4}$	$a_{i,gA}$	$b_{i,gA}$	$a_{i,gB}$	$b_{i,gB}$	$a_{i,R}$	$b_{i,R}$	H	G	Decision
F_{01}	S_5	0.01	0.94	0.13	0.88	0.14	1.00	0.17	0.91	0.13	0.88	0.17	0.91	0.13	0.91	−0.78	0.04	UNVIABLE
F_{02}	S_5	0.00	0.95	0.15	0.75	0.15	0.85	0.25	1.00	0.15	0.75	0.25	0.85	0.15	0.85	−0.7	0.00	UNVIABLE
F_{03}	S_5	0.05	0.98	0.17	0.83	0.18	0.02	0.21	0.95	0.17	0.83	0.21	0.02	0.17	0.83	−0.66	0.00	UNVIABLE
F_{04}	S_5	0.08	1.00	0.20	0.86	0.21	0.05	0.24	0.98	0.20	0.86	0.24	0.05	0.20	0.86	−0.66	0.06	UNVIABLE
F_{05}	S_5	0.00	1.00	0.10	0.80	0.90	0.08	1.00	0.15	0.10	0.80	1.00	0.08	0.10	0.80	−0.70	−0.10	UNVIABLE
F_{06}	S_5	0.10	0.80	0.20	0.90	0.13	1.00	0.00	1.00	0.20	0.80	0.13	1.00	0.13	1.00	−0.87	0.13	UNVIABLE
F_{07}	S_5	0.05	0.88	0.15	0.85	0.12	1.00	0.00	0.95	0.15	0.85	0.12	0.95	0.12	0.95	−0.83	0.07	UNVIABLE
F_{08}	S_5	0.08	0.83	0.18	0.95	0.21	0.95	0.05	0.98	0.18	0.83	0.21	0.95	0.18	0.95	−0.77	0.13	UNVIABLE
F_{09}	S_5	0.10	0.86	0.15	0.93	0.24	0.98	0.08	1.00	0.15	0.86	0.24	0.98	0.15	0.98	−0.83	0.13	UNVIABLE
F_{10}	S_5	0.13	0.79	0.14	0.90	0.17	0.91	0.01	0.94	0.14	0.79	0.17	0.91	0.14	0.91	−0.77	0.05	UNVIABLE
Baricenter W: averages of the resultant degrees														0.147	0.904	−0.757	0.051	UNVIABLE

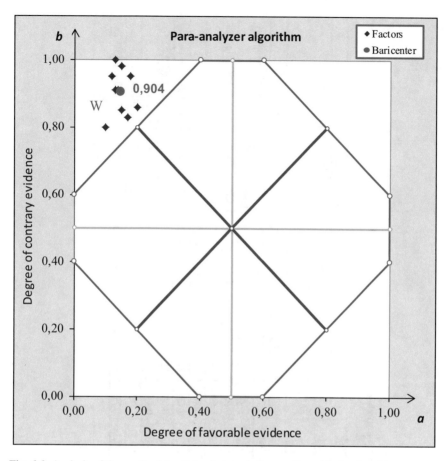

Fig. 6.6 Analysis of the result when all the factors are very unfavorable (section S_5)

Table 6.10 Application of the PDM when five factors are very favorable (section S_1) and five one are favorable only (section S_2), with level of requirement 0.60

1	2	3	4	5	6	7	8	9	10	11	12	13	14	15	16	17	18	19
F_i	S_j	Group A				Group B				A		B		A MIN B		Conclusions		Level of requirement = 0.60
		E_1		E_2		E_3		E_4		E_1 MAX E_2		E_3 MAX E_4						
		$a_{i,1}$	$b_{i,1}$	$a_{i,2}$	$b_{i,2}$	$a_{i,3}$	$b_{i,3}$	$a_{i,4}$	$b_{i,4}$	$a_{i,gA}$	$b_{i,gA}$	$a_{i,gB}$	$b_{i,gB}$	$a_{i,R}$	$b_{i,R}$	H	G	Decision
F_{01}	S_2	0.63	0.19	0.79	0.23	0.73	0.14	0.59	0.24	0.79	0.19	0.73	0.19	0.54	0.19	0.54	−0.08	NOT CONCLUSIVE
F_{02}	S_1	1.00	0.05	0.95	0.15	1.00	0.10	0.85	0.00	1.00	0.05	1.00	0.05	0.95	0.05	0.95	0.05	VIABLE
F_{03}	S_2	0.67	0.23	0.83	0.27	0.77	0.18	0.63	0.28	0.83	0.23	0.77	0.23	0.54	0.23	0.54	0.00	NOT CONCLUSIVE
F_{04}	S_1	0.95	0.11	1.00	0.21	0.91	0.15	0.85	0.10	1.00	0.11	0.91	0.11	0.80	0.11	0.80	0.02	VIABLE
F_{05}	S_2	0.70	0.20	0.80	0.30	0.80	0.20	0.70	0.30	0.80	0.20	0.80	0.20	0.60	0.20	0.60	0.00	VIABLE
F_{06}	S_1	0.90	0.10	1.00	0.10	0.90	0.00	1.00	0.00	1.00	0.10	2.00	0.10	0.90	0.10	0.90	0.10	VIABLE
F_{07}	S_2	0.85	0.25	0.85	0.30	0.73	0.35	0.75	0.25	0.85	0.25	0.75	0.25	0.50	0.25	0.50	0.00	NOT CONCLUSIVE
F_{08}	S_1	0.98	0.18	0.88	0.12	0.82	0.07	0.92	0.08	0.98	0.12	0.92	0.12	0.80	0.22	0.80	0.04	VIABLE
F_{09}	S_1	1.00	0.21	0.71	0.15	0.85	0.10	0.95	0.11	1.00	0.15	0.95	0.15	0.80	0.15	0.80	0.10	VIABLE
F_{10}	S_2	0.79	0.23	0.73	0.14	0.59	0.24	0.63	0.19	0.79	0.14	0.93	0.19	0.44	0.19	0.44	−0.18	NOT CONCLUSIVE
Baricenter W: averages of the resultant degrees														0.846	0.159	0.687	0.005	VIABLE

Result: **VIABLE**

(a)

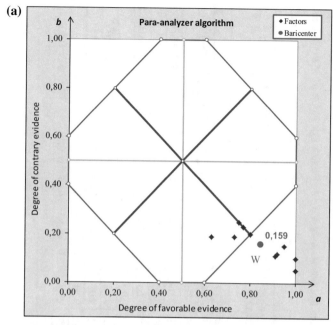

(b)

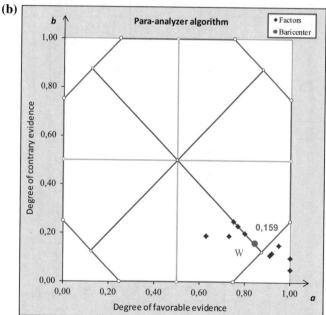

Fig. 6.7 a Analysis of the results by the para-analyzer algorithm in case five factors are very favorable (S_1), five are only favorable (S_2) and the requirement level is equal to **0.60**. Result: **feasible**. **b** Analysis of the results by the para-analyzer algorithm in case five factors are very favorable (S_1), five are only favorable (S_2) and the requirement level is equal to **0.75**. Result: **inconclusive**

Table 6.11 Application of the PDM when five factors are very favorable (section S_1) and five one are favorable only (section S_2), with level of requirement 0.75

1	2	3	4	5	6	7	8	9	10	11	12	13	14	15	16	17	18	19
F_i	S_j	Group A				Group B				A		B		A MIN B		Conclusions		Level of requirement = 0.75
		E_1		E_2		E_3		E_4		E_1 MAX E_2		E_3 MAX E_4						
		$a_{i,1}$	$b_{i,1}$	$a_{i,2}$	$b_{i,2}$	$a_{i,3}$	$b_{i,3}$	$a_{i,4}$	$b_{i,4}$	$a_{i,gA}$	$b_{i,gA}$	$a_{i,gB}$	$b_{i,gB}$	$a_{i,R}$	$b_{i,R}$	H	G	Decision
F_{01}	S_2	0.63	0.19	0.79	0.23	0.73	0.14	0.59	0.24	0.79	0.19	0.73	0.14	0.73	0.19	0.54	−0.08	NOT CONCLUSIVE
F_{02}	S_1	1.00	0.05	0.95	0.15	1.00	0.10	0.85	0.00	1.00	0.05	1.00	0.00	1.00	0.05	0.95	0.05	VIABLE
F_{03}	S_2	0.67	0.23	0.83	0.27	0.77	0.18	0.63	0.28	0.83	0.23	0.77	0.18	0.77	0.23	0.54	0.00	NOT CONCLUSIVE
F_{04}	S_1	0.95	0.11	1.00	0.21	0.91	0.15	0.85	0.10	1.00	0.11	0.91	0.10	0.91	0.11	0.80	0.02	VIABLE
F_{05}	S_2	0.70	0.20	0.80	0.30	0.80	0.20	0.70	0.30	0.80	0.20	0.80	0.20	0.80	0.20	0.60	0.00	NOT CONCLUSIVE
F_{06}	S_1	0.90	0.10	1.00	0.10	0.90	0.00	1.00	0.00	1.00	0.10	1.00	0.00	1.00	0.10	0.90	0.10	VIABLE
F_{07}	S_2	0.85	0.25	0.85	0.30	0.73	0.35	0.75	0.25	0.85	0.25	0.75	0.25	0.75	0.25	0.50	0.00	NOT CONCLUSIVE
F_{08}	S_1	0.98	0.18	0.88	0.12	0.82	0.07	0.92	0.08	0.98	0.12	0.92	0.07	0.92	0.12	0.80	0.04	VIABLE
F_{09}	S_1	1.00	0.21	0.91	0.15	0.85	0.10	0.95	0.11	1.00	0.15	0.95	0.10	0.95	0.15	0.80	0.10	VIABLE
F_{10}	S_2	0.79	0.23	0.73	0.14	0.59	0.24	0.63	0.19	0.79	0.14	0.63	0.19	0.63	0.19	0.44	−0.18	NOT CONCLUSIVE
Baricenter W: averages of the resultant degrees														0.846	0.159	0.687	0.005	

Result: **NOT CONCLUSIVE**

6.3   Evaluation of a Factory Project

In this example, the PDM will be applied in the assessment of the project P of a factory, a problem which engineers, consultants or business people continuously face. The idea is to analyze the feasibility of implementation of the project of a factory.

6.3.1   Establishment of the Requirement Level

The first mission of the KE is to establish the requirement level of the analysis for the decision making. In this application, it will be fixed at 0.65. With that, the para-analyzer algorithm and the decision rule are already determined (Fig. 6.8).

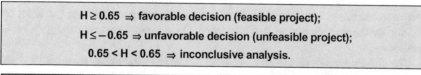

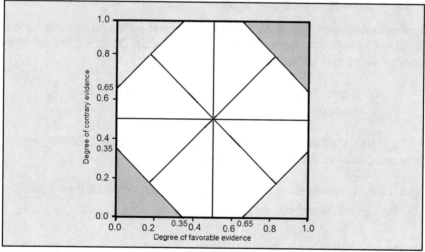

Fig. 6.8 Decision rule and para-analyzer algorithm for requirement level **0.65**

6.3.2   Selection of Factors and Establishment of the Sections

Eight factors (F_1 to F_8) that influence the decision to implement or not the project of a factory, that is, which influence the project feasibility, will be chosen. For each one of these factors, three sections (S_1 to S_3) will be established, so that S_1 represents a **favorable** situation; S_2, an **indifferent** case; and S_3, an **unfavorable** situation to the implementation of the project.

In the choice of the influence factors, we must seek to encompass the different aspects involved in the structure of a project: economic (which involve market, location and scale), technical (production process, project engineering, physical the arrangement of the equipment, etc.), financial (composition of the capital, loans, working capital, etc.), administrative (organizational structure of the implementation moreover, operation), legal (contracts with suppliers, purchase of technology and patents),environmental and accounting aspects [109].

So, the choice of the factors and the characterization of the sections depend on the project being evaluated, on market analyses, on economic studies and other elements. In this study, this characterization was performed without the strict utilization of these elements, as it is a theoretical example for the presentation of the method.

The chosen factors (F_i), with $1 \leq i \leq 8$, and the established sections (S_j), with $1 \leq j \leq 3$, are the ones presented below.

F_1: **Production capacity of the factory**—Measured by the comparison between the forecast production for the factory and the average M of production of the already existent similar factories.

S_1 : higher than 1.2M; S_2 : a different situation from S_1 and S_3;
S_3 : lower than 0.8M.

F_2: **Equipment selection**—Translated by the characteristics: flexibility, productivity and quality of the chosen equipment.

S_1 : the three characteristics are high; S_2 : a different situation from S_1 and S_3;
S_3 : the three characteristics are low.

F_3: **Factory layout**—Translated by the characteristics: ease of material entry, proper arrangement of the equipment for production flow and ease of production outflow.

S_1 : the three characteristics are high; S_2 : a different situation from S_1 and S_3;
S_3 : the three characteristics are low.

F$_4$: Location—Translated by the proximity to the following elements: material supplier center, consumer center, good roads and inexpensive means of transportation (railway or waterway).

S$_1$: at least three of these elements are very near;
S$_2$: a different situation from S$_1$ and S$_3$;
S$_3$: at least three of these elements are **not** very near.

F$_5$: Organization—Translated by the adequacy of the following support systems: quality control, maintenance, packaging system and product dispatch logistics.

S$_1$: at least three of these systems are very adequate;
S$_2$: a different situation from S$_1$ and S$_3$;
S$_3$: at least three of these systems are **not** very adequate.

F$_6$: Internal area availability—Measured by the percentage of free area for eventual stocking or the implementation of new departments.

$$S_1 : \text{more than } 50\%; \quad S_2 : \text{a different situation from } S_1 \text{ and } S_3;$$
$$S_3 : \text{less than } 20\%.$$

F$_7$: Possibility of expansion—Measured by the ratio between the total property area and the area occupied by the project.

$$S_1 : \text{higher than } 3; \quad S_2 : \text{a different situation from } S_1 \text{ and } S_3;$$
$$S_3 : \text{lower than } 2.$$

F$_8$: Process flexibility—Translated by the capacity of adaptation for the production of different products.

$$S_1 : \text{high capacity}; \quad S_2 : \text{medium capacity}; \quad S_3 : \text{low capacity.}$$

6.3.3   Construction of the Database

After the factors have been chosen and the sections established, by means of specialists (or utilizing statistical data), the degree of favorable evidence (or belief) ($a_{i,j,k}$) and degree of contrary evidence (or disbelief) ($b_{i,j,k}$) are attributed to the project success, for each one of the factors, in each one of the sections, as well as the weights for each one of the factors. In this example, the option is for the use of specialists.

According to criteria established by the KE, the following specialists were chosen: E$_1$: production engineer; E$_2$: industrial administrator; E$_3$: process engineer (mechanical or chemical, or another one, depending on the factory); and E$_4$: product engineer.

The average weights of the factors, as well as the degrees of the favorable evidence and contrary evidence attributed by the specialists to the factors, in the conditions of the established sections, are in Table 6.12, which constitutes the **database**.

Table 6.12 Database: the average weights of the factors and the degrees of the favorable evidence and contrary evidence attributed by the specialists to the factors, in each one of the sections

Factor	Weight	Section	Spec 1		Spec 2		Spec 3		Spec 4	
F_i	P_i	S_j	$a_{i,1}$	$b_{i,1}$	$a_{i,2}$	$b_{i,2}$	$a_{i,3}$	$b_{i,3}$	$a_{i,4}$	$b_{i,4}$
F1	1	S1	1.0	0.0	0.9	0.1	1.0	0.2	0.8	0.3
		S2	0.7	0.4	0.6	0.4	0.6	0.6	0.5	0.6
		S3	0.3	1.0	0.3	1.0	0.2	0.8	0.2	1.0
F2	1	S1	0.9	0.2	1.0	0.2	0.9	0.1	0.8	0.1
		S2	0.6	0.5	0.6	0.6	0.4	0.4	0.5	0.4
		S3	0.3	0.9	0.2	0.8	0.1	0.8	0.0	0.9
F3	1	S1	0.9	0.2	0.8	0.2	0.8	0.0	0.7	0.2
		S2	0.6	0.4	0.4	0.4	0.6	0.5	0.5	0.5
		S3	0.3	1.0	0.0	1.0	0.3	1.0	0.1	0.9
F4	3	S1	1.0	0.2	0.8	0.0	1.0	0.2	0.9	0.4
		S2	0.5	0.6	0.6	0.6	0.6	0.4	0.5	0.6
		S3	0.1	1.0	0.2	1.0	0.2	1.0	0.01	0.9
F5	1	S1	0.9	0.9	1.0	0.8	1.0	0.1	0.2	0.9
		S2	0.4	0.5	0.6	0.3	0.7	0.3	0.5	0.6
		S3	0.1	0.8	1.0	0.2	0.9	0.3	0.8	0.3
F6	1	S1	1.0	0.1	1.0	0.1	1.0	0.0	0.1	0.8
		S2	0.6	0.5	0.7	0.3	0.7	0.4	0.6	0.4
		S3	0.3	1.0	0.2	0.9	0.3	0.9	0.0	0.9
F7	2	S1	1.0	0.2	1.0	0.0	0.9	0.2	1.0	0.2
		S2	0.6	0.5	0.3	0.4	0.6	0.5	0.5	0.6
		S3	0.1	1.0	0.3	0.9	0.3	0.7	0.0	0.9
F8	2	S1	1.0	0.2	0.9	0.2	0.9	0.1	0.8	0.2
		S2	0.7	0.3	0.6	0.5	0.5	0.4	0.5	0.6
		S3	0.0	0.9	0.3	0.7	0.3	0.8	0.2	0.9

6.3.4   Field Research and Obtention of the Results

We must research about the project P, to verify in which section each factor is. That is, the researchers must verify, for each one of the factors F_i ($1 \leq i \leq 8$), in which section S_j ($1 \leq j \leq 3$) the project P is. With the sections S_{pj} found in the research, column 3 of Table 6.13 is filled out. Having these results, the CP of the PDM extracts from the database (Table 6.12), besides the average weights of the factors (column 2), the experts' opinions concerning the feasibility of project P in the conditions of the factors F_i, translated by the researched sections. These opinions, translated by the degrees of evidence, favorable and contrary, are placed in columns 4–11 of Table 6.13.

Then, the CP of the PDM applies the maximization (**MAX** operator) and minimization (**MIN** operator) techniques of the lattice associated to logic Eτ. In this application, it is convenient that the groups are constituted observing the specialists' majors. That is almost always a choice of the KE.

Table 6.13 Calculation table of the PDM: factors (column 1), weights (2), researched sections (3), degrees of favorable and contratry evidence (4–11), MAX and MIN operators application (12 and 15), calculations (16 and 17) and result analysis (18)

1	2	3	4	5	6	7	8	9	10	11	12	13	14	15	16	17	18
Factor	Weight	Section	Group A		Group B		Group C		Spec 4		Group C		MIN application		Level of Req.		0.65
			Spec 1		Spec 2		Spec 3				MAX [E3, E4]		MIN [A, B, C]		Conclusions		
F_i	P_i	S_j	$a_{i,1}$	$b_{i,1}$	$a_{i,2}$	$b_{i,2}$	$a_{i,3}$	$b_{i,3}$	$a_{i,4}$	$b_{i,4}$	$a_{i,C}$	$b_{i,C}$	$a_{i,R}$	$b_{i,R}$	H	G	Decision
F1	1	S3	0.3	1.0	0.3	1.0	0.2	0.8	0.2	1.0	0.2	0.8	0.2	1.0	−0.8	0.2	UNVIABLE
F2	1	S1	0.9	0.0	1.0	0.1	0.9	0.1	0.8	0.0	0.9	0.0	0.9	0.1	0.8	0.0	VIABLE
F3	1	S2	0.6	0.4	0.4	0.4	0.6	0.5	0.5	0.5	0.6	0.5	0.4	0.5	−0.1	−0.1	NON CONCLUSIVE
F4	3	S3	0.1	1.0	0.2	1.0	0.2	1.0	0.0	0.9	0.2	0.9	0.1	1.0	−0.9	0.1	UNVIABLE
F5	1	S1	0.9	0.9	1.0	0.8	1.0	0.1	0.2	0.9	1.0	0.1	0.9	0.9	0.0	0.8	NON CONCLUSIVE
F6	1	S2	0.6	0.5	0.7	0.3	0.7	0.4	0.6	0.4	0.7	0.4	0.6	0.5	0.1	0.1	NON CONCLUSIVE
F7	2	S3	0.1	1.0	0.3	0.9	0.3	0.7	0.0	0.9	0.3	0.7	0.1	1.0	−0.9	0.1	UNVIABLE
F8	2	S1	1.0	0.0	0.9	0.0	0.9	0.2	0.8	0.0	0.9	0.0	0.9	0.0	0.9	−0.1	VIABLE
Baricenter W: weighted average of the resultant degrees													0.44	0.67	−0.23	0.11	NON CONCLUSIVE

Suppose that, in the utilized specialist's group, the KE considers that the opinions of experts 1 and 2 are indispensable, but that, among experts 3 and 4, one favorable opinion is sufficient. Thus, the formation of the groups is group A—production engineer (E_1); group B—industrial administrator (E_2); and group C—process engineer (E_3) with product engineer (E_4). This way, for the application of the maximization (**MAX**) and minimization (**MIN**) techniques to these experts' opinions, we must do:

$$\mathbf{MIN}\{(E_1), (E_2), \mathbf{MAX}[(E_3), (E_4)]\} \text{ or } \mathbf{MIN}\{G_A, G_B, G_C\}$$

that is, first, the **MAX** operator is applied first only inside group C (intragroup) and, then, the **MIN** operator is applied to groups A, B and C (intergroup).

In Table 6.13, the results of the application of the **MAX** operator to group C (intragroup) are in columns 12 to 13. The results of the application of the **MIN** operator among groups A, B and C (intergroup), appear in columns 14 and 15. This way, we obtain, for each factor, in the conditions of the section found in the research, the degrees of belief ($a_{i,R}$) and disbelief ($b_{i,R}$), resultant from the combination of the experts' opinions. It is appropriate to remember that all the previously described operations are conducted by the CP of the PDM.

6.3.5   Analysis of the Results and Final Decision

Using the CP of the PDM, the calculations are performed, as well as the analysis of the results by the application of the decision rule, which appears in column 18 of Table 6.13. That enables us to say that it is the influence of each factor (F_1 to F_8) on the decision of feasibility of project P, as well as the joint influence of all the factors, through the center of gravity **W**, in the conditions characterized by the sections obtained in the research. The observation of column 18 shows that two factors (F_2 and F_8) recommend the execution of project P, at the requirement level of 0.65; that three factors (F_1, F_4 and F_7) recommend the non-execution of project P, at the requirement level of 0.65; and that the other factors (F_3, F_5 and F_6) are inconclusive. It must be observed that factor F_5, besides being inconclusive, presents a high degree of uncertainty (contradiction) ($G = 0.80$), showing that, in relation to this factor in the researched section (S_1), there is great inconsistency among the specialists' opinions.

In the analysis of the joint influence of all the factors, which is translated by the center of gravity **W**, the last line of column 18 of Table 6.13 shows that the global result is inconclusive, that is, the analysis does not enable us to infer for the feasibility of the project, nor for its unfeasibility. In this case, it is recommended to give up the project or conduct further analyses, so that new evidences may emerge.

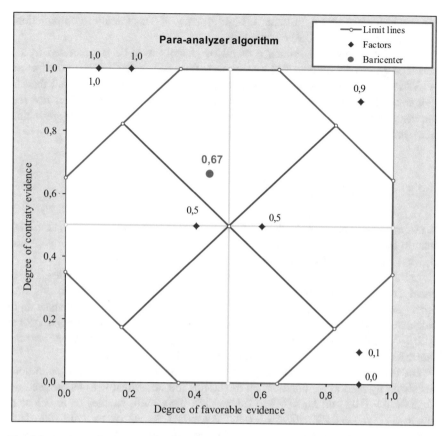

Fig. 6.9 Analysis of the results by the para-analyzer device, with requirement level equal to **0.65**

The analysis of the result, as seen in the previous examples, may also be performed by the para-analyzer algorithm, plotting the resultant degrees of favorable evidence and contrary evidence for the factors, and of the center of gravity, in lattice τ, according to Fig. 6.9.

From the observation of Fig. 6.9, we observe that two factors (in the figure, represented by diamonds) belong to the truth region, recommending the execution of the project; three factors belong to the falsity region, recommending, then, the non-execution of project P [in Fig. 6.9, only two points appear, because $F_4 \equiv F_7 = (0.10; 1.00)$]; and that the other factors belong to other regions, being, therefore, inconclusive, everything at the requirement level of 0.65.

It must be observed that factor F_5 belongs to the inconsistency region, showing that the specialists' opinions, in relation to this factor in the researched section (S_1), are contradictory (they present a high degree of uncertainty (contradiction), $G = a + b - 1 = 0.9 + 0.9 - 1 = 0.80$).

The joint (combined) influence of all the factors may be summarized by the center of gravity **W** (represented by a circle) of the points that represent the factors. As **W** is in the region of quasi-falsity, tending to inconsistency, it is said that the global result of the analysis is **inconclusive**. That is, the analysis does not recommend project P, but does not exclude this possibility either. It only suggests that, new analyses are conducted, in an attempt to increase the evidences.

6.3.6 Reliability of the PDM

A test of the reliability of the decision method (PDM) may be conducted, admitting that, in the field research, all the factors are in the conditions translated by section S_1, that is, all the factors are **favorable** to the feasibility of a project.

In this case, evidently, it is not a surprise that the method application leads us to conclude for the feasibility of the project. In fact, applying the PDM to this situation, we obtain $a_W = 0.89$ and $b_W = 0.20$ (Table 6.14). This result enables us to calculate the degree of certainty of the center of gravity $H_W = a_W - b_W = 0.89 - 0.20 = 0.69$. As $0.69 \geq 0.65$, the decision rule (Fig. 6.8) enables us to infer for the **feasibility** of the project, at the requirement level of 0.65 (Fig. 6.10a).

On the contrary, if in another scenario, all the factors are in the conditions translated by section S_3, that is, if all the factors are **unfavorable** to the project, we obtain $a_W = 0.09$ and $b_W = 0.96$ (Table 6.15). This result enables us to calculate $H_W = a_W - b_W = 0.09 - 0.96 = -0.87$. As $-0.87 \geq -0.65$, applying the decision rule, we conclude, in this scenario, for the **unfeasibility** of the other project, at the requirement level of 0.65 (Fig. 6.10b).

6.3.7 Influence of the Requirement Level

To verify the influence of the requirement level in the decision, a situation was analyzed, in which five factors (F_2, F_3, F_4, F_7 and F_8) are favorable to the project (section S_1) and other three (F_1, F_5, and F_6) are indifferent (section S_2). In this situation, the evaluation of the project results as **inconclusive,** at the requirement level of 0.65 (Fig. 6.11a), but it certifies its **feasibility** at the requirement level of 0.50 (Fig. 6.11b).

Table 6.14 Analysis done for all factors in the more favorable conditions (section S₁)

1	2	3	4	5	6	7	8	9	10	11	12	13	14	15	16	17	18
Factor	Weight	Section	Group A		Group B		Group C		Spec 4		Group C		MIN application		Level of Req.		Conclusions
			Spec 1		Spec 2		Spec 3						MAX [E3, E4]		MIN [A, B, C]		
F_i	P_i	S_j	$a_{i,1}$	$b_{i,1}$	$a_{i,2}$	$b_{i,2}$	$a_{i,3}$	$b_{i,3}$	$a_{i,4}$	$b_{i,4}$	$a_{i,C}$	$b_{i,C}$	$a_{i,R}$	$b_{i,R}$	H	G	Decision
F1	1	S1	1.0	0.0	0.9	0.1	1.0	0.2	0.8	0.3	1.0	0.2	0.9	0.2	0.7	0.1	VIABLE
F2	1	S1	0.9	0.0	1.0	0.1	0.9	0.1	0.8	**0.0**	0.9	0.0	0.9	0.1	0.8	0.0	VIABLE
F3	1	S1	0.9	0.0	0.8	0.1	0.8	0.0	0.9	0.1	0.9	0.0	0.8	0.1	0.7	−0.1	VIABLE
F4	3	S1	1.0	0.2	0.8	0.0	1.0	0.2	0.9	0.4	1.0	0.2	0.8	0.2	0.6	0.0	NOT CONCLUSIVE
F5	1	S1	0.9	0.9	1.0	0.8	1.0	0.1	0.2	0.9	1.0	0.1	0.9	0.9	0.0	0.8	NOT CONCLUSIVE
F6	1	S1	1.0	0.1	1.0	0.1	1.0	0.0	0.1	0.8	1.0	0.0	1.0	0.1	0.9	0.1	VIABLE
F7	2	S1	1.0	0.2	1.0	0.0	0.9	0.2	1.0	0.2	1.0	0.2	1.0	0.2	0.8	0.2	VIABLE
F8	2	S1	1.0	0.0	0.9	0.0	0.9	0.2	0.8	0.0	0.9	0.0	0.9	0.0	0.9	−0.1	VIABLE
Baricenter W: weighted average of the resultant degrees													0.89	0.20	0.69	0.09	0.65

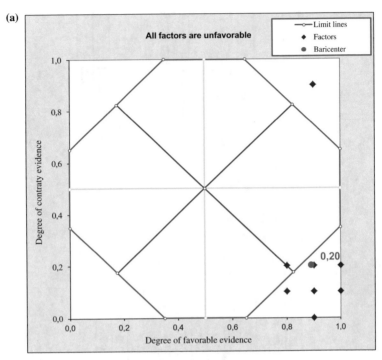

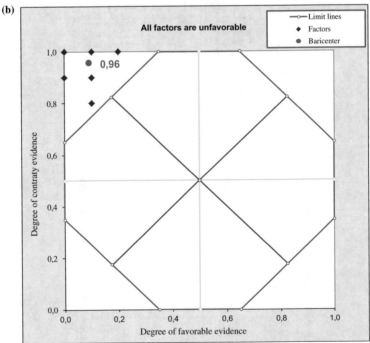

Fig. 6.10 a, b Analysis of the project in other scenarios: when all the factors are favorable (**feasible** to the requirement level **0.65**) and when all the factors are unfavorable (**unfeasible** to the requirement level **0.65**)

Table 6.15 Analysis done for all factors in the more unfavorable conditions (section S_3)

1	2	3	4	5	6	7	8	9	10	11	12	13	14	15	16	17	18
Factor	Weight	Section	Group A		Group B		Group C				Group C		MIN application	MIN [A, B, C]	Level of Req.		0.65
			Spec 1		Spec 2		Spec 3		Spec 4		MAX [E3, E4]				Conclusions		
F_i	P_i	S_j	$a_{i,1}$	$b_{i,1}$	$a_{i,2}$	$b_{i,2}$	$a_{i,3}$	$b_{i,3}$	$a_{i,4}$	$b_{i,4}$	$a_{i,C}$	$b_{i,C}$	$a_{i,R}$	$b_{i,R}$	H	G	Decision
F1	1	S3	0.3	1.0	0.3	1.0	0.2	0.8	0.2	1.0	0.2	0.8	0.2	1.0	−0.8	0.2	UNVIABLE
F2	1	S3	0.3	0.9	0.2	0.8	0.1	0.8	0.0	0.9	0.1	0.8	0.1	0.9	−0.8	0.0	UNVIABLE
F3	1	S3	0.3	1.0	0.0	1.0	0.3	1.0	0.1	0.9	0.3	0.9	0.0	1.0	−1.0	0.0	UNVIABLE
F4	3	S3	0.1	1.0	0.2	1.0	0.2	1.0	0.0	0.9	0.2	0.9	0.1	1.0	−0.9	0.1	UNVIABLE
F5	1	S3	0.1	0.8	1.0	0.2	0.9	0.3	0.8	0.3	0.9	0.3	0.1	0.8	−0.7	−0.1	UNVIABLE
F6	1	S3	0.3	1.0	0.2	0.9	0.3	0.9	0.0	0.9	0.3	0.9	0.2	1.0	−0.8	0.2	UNVIABLE
F7	2	S3	0.1	1.0	0.3	0.9	0.3	0.7	0.0	0.9	0.3	0.7	0.1	1.0	−0.9	0.1	UNVIABLE
F8	2	S3	0.0	0.9	0.3	0.7	0.3	0.8	0.2	0.9	0.3	0.8	0.0	0.9	−0.9	−0.1	UNVIABLE
Baricenter W: weighted average of the resultant degrees													0.09	0.96	−0.87	0.05	UNVIABLE

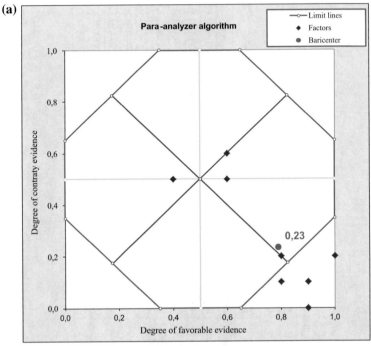

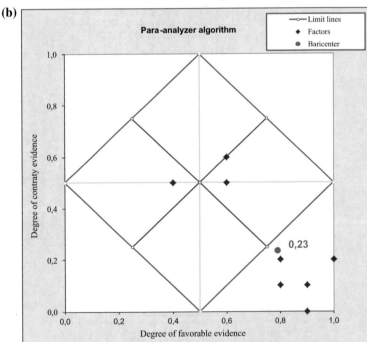

Fig. 6.11 a, b Analysis of the result when five factors are favorable (section S_1,) and three are indifferent (section S_2). Result: **inconclusive**, at the requirement level **0.65**; **feasible**, at the requirement level **0.50**

6.4 Feasibility Analysis of the Implementation of a Manufacturing System that Utilizes Advanced Technologies

At the moment when the machinery of a company is obsolete and needs to be replaced, the businesspeople or the administrators have two options: (1) maintain the manufacturing system of the production process and only replace the old machines for new machines, but equal or (2) make an innovation, replacing it for a more modern one, with the introduction of advanced technologies (new machines, new techniques, new processes, etc.).

If the option is introducing technological innovation, the doubt remains: which technological innovation is the most adequate? There are several options of manufacturing systems that utilize advanced technologies to be adopted, and each one of them may present advantages or disadvantages in relation to the previous system, traditional. These benefits or disadvantages are connected to strategic factors and economic-operational factors, some of them of qualitative character, and others of quantitative character. These factors, in turn, are related to the amount of the capital to be invested and to the operational and financial results from these investments [18].

Then, the intention in this example is, analyzing the influence of these factors in a combined manner, to verify: (1) to verify if there is an advantage or not in replacing the old manufacturing system with the traditional technology for a modern manufacturing system, provided with advanced technologies and, also, (2) which of the new systems is the most adequate.

Currently, several manufacturing systems with technological innovations may be introduced in the everyday functioning of a factory, and among them, we highlight: *CAD/CAM*—Computer-Automated Design and Manufacture; *GT/CM*—Group Technology and Cellular Manufacture; *RE*—Robotics Equipment; *FMS*—Flexible Manufacturing Systems; *AA*—Automated Assembly; *CIM*—Computer-Integrated Manufacture [22].

The initials CAD/CAM is more commonly understood as "project aided by computer and manufacture aided by computer" or project and manufacture aided by computer [57].

However, there are several attributes (factors or indicators), whose performances influence the result of the implementation of these innovations, in order to make them advantageous or not in relation to the traditional process that was previously utilized. The feasibility or not of the replacement of the old system for the new system will be decided by means of the comparative analysis of the performance of these attributes in the new and in the old system.

Below is a list of these factors, separated by class:

- factors related to the company's strategic goals: technological reputation, market share (compartment), competitive position and product innovation;
- qualitative or quantitative factors, of economic or operational character, related to the amount of the capital to be invested: product heterogeneity, number of

produced pieces, payback period, net present value (NPV), future operational costs, residual amounts, service life, real time measurements, delivery, product reliability, answer time, direct labor savings, creation funding, space in the shop floor, additional indirect labor, scraps, warranty claim (or right), replacement time, preparations, remains, reprocessing cost, etc. [22].

The purpose of this item is to compare one of the mentioned new manufacturing systems, provided with advanced technologies, to the traditional one, to know which one is better, that is, which one conducts the company to a better economic result. For each analysis, the most significant influence factors on the performance of the manufacturing system will be selected, to compare the new adopted system, taking the old system into account. Utilizing techniques of the paraconsistent annotated logic to analyze the combined influence of all the factors, we may conclude if the introduction of the new system with advanced technologies produces better result for the company than the old one.

6.4.1 Performance Coefficient of a New Manufacturing System Compared to the Old One, for a Certain Influence Factor

First of all, a number will be defined, which translates the performance of a new manufacturing system (N), utilizing advanced technologies, compared to the old system (A), for a certain indicator (I) or influence factor. Then, the stages of the PDM will be analyzed.

Consider I_0 and I the values of a performance indicator of the company in the old manufacturing system and in the new one, respectively. For this indicator, the performance coefficient (CD) of the new manufacturing system compared to the old one will be defined, as follows:

$$CD_{N,A}(I) = 1 \pm (\Delta I/I_0), \text{ where } \Delta I = I - I_0. \qquad (6.4.1)$$

The signal $\pm$ must be used as follows: if the performance (**D**) of the system improves when I increases, we use the signal +; if the pe (**D**) of the system improves when I decreases, we use the signal $-$. That is, we use the signal + when **D** is an increasing function of I, and the signal $-$ when **D** is a decreasing function of I.

We may say, for instance, that the performance (**D**) of a new productive system in relation to the old one is an increasing function of the gross revenue (R). In fact, it is accepted that the performance of the production system becomes better when it manages to increase the company's revenue (R). Thus, if the utilized indicator is the revenue, we must use the signal +. Numerically, if, with the introduction of the new system, the income grows from $I_0 = \$10,000$ to $I = \$12,000$, then the performance coefficient, in relation to the revenue, will be: $CD_{NA}(R) = 1 + (+2,000/10,000) = 1.20$; if the revenue decreases from $I_0 = \$10,000$ to $I = \$8,000$, then the performance coefficient, in relation to the revenue, will be: $CD_{N,A}$, $(R) = 1 + (-2,000/10,000) = 0.80$.

Analogously, we may say that the performance (**D**) of a new productive system in relation to the old one is a decreasing function of the production cost (C). Thus, if the utilized indicator is the production cost, we must use the signal $-$. If, with the introduction of the new system, the company's production cost increases from $I_0 = \$10,000$ to $I = \$13,000$, then the performance coefficient will be: CD_{NA} $(C) = 1 - (+3,000/10,000) = 0.70$. That is, the production cost increases and CD is lower than 1 (the performance becomes worse). If the company's production cost decreases from $I_0 = \$10,000$ to $I = \$7,000$, then the performance coefficient will be: CD_{NA} $(C) = 1 - (-3,000/10,000) = 1.30$. That is, the production cost decreases and CD is lower than 1 (the performance improves).

These examples enable us to observe that, if $CD_N > 1$, the new production system (N) improves the company's performance in relation to the old one (A); if $CD_N < 1$, it worsens.

6.4.2 Establishment of the Requirement Level

For the decision making, the requirement level of 0.75 will be established. That means that the decision will be made if $|H| \geq 0.75$, that is, the segments defined by the condition $|H| = 0.75$ are being adopted as limit lines of truth and falsity. Therefore, the decision will only be favorable if the degree of favorable evidence outweighs the degree of contrary evidence in at least 0.75. That is a high requirement level.

Thus, the para-analyzer algorithm and the decision rule are already determined (Fig. 6.12).

6.4.3 Identifying the Influence Factors (Attributes or Indicators)

Below, we will present a list of influence factors that may (or may not) be utilized in the feasibility analysis of a new manufacturing system with advanced technologies; the use or not of the factor depends on the system under analysis. For some systems, the factor may be important; for others, not. Then, the importance of the factor is relative. The most important factors will be utilized in the analysis; the least important ones, or the ones without any importance, will not be used.

Two classes of factors are emphasized [22]. Firstly, the factors related to the company's strategic goals. These are the non-measurable factors, almost intangible, so that they may only be framed in a section by experienced specialists in the subject.

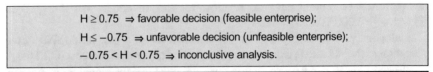

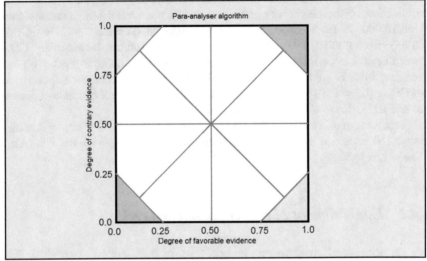

Fig. 6.12 Decision rule and para-analyzer algorithm for requirement level **0.75**

F_{01}—technological reputation
F_{02}—market share occupied by the company
F_{03}—competitive position of the company inside the market occupied by it
F_{04}—innovation of the product produced by the company

Second, we highlight the factors related to the company's operational and economic results. Among these factors, some are measurable and others are not. The first ones may be framed in the section by quantitative criteria, but the second ones may only be framed by qualitative criteria, by means of experienced and specialized people in the subject.

F_{05}—Total investment	F_{06}—Total expenses
F_{07}—Net present value (NPV)	F_{08}—Payback period
F_{09}—Residual amounts	F_{10}—Creation funding
F_{11}—Product heterogeneity	F_{12}—Product reliability
F_{13}—System service life	F_{14}—System flexibility
F_{15}—Future operational costs	F_{16}—Direct labor costs
F_{17}—Reprocessing cost	F_{18}—Additional indirect labor costs
F_{19}—Material cost	F_{20}—Capital investment cost
F_{21}—Real time measurements	F_{22}—Replacement time
F_{23}—Delivery time	F_{24}—Answer time
F_{25}—Preparation time	F_{26}—Machine utilization time
F_{27}—Waiting time	F_{28}—Space in the shop floor
F_{29}—Number of produced pieces	F_{30}—Warranty claim (or right)
F_{31}—Scraps	F_{32}—Remains

6.4.4 Establishing the Sections for the Influence Factors

For each indicator, five sections will be established, S_j, with $1 \leq j \leq 5$, so that S_1 represents a much better situation of the new system with advanced technologies when compared to the old system; S_2 represents a better situation; S_3, an indifferent situation; S_4, a worse situation; and S_5 represents a much worse situation of the new system when compared to the old one.

It will be said that the situation of the new system with advanced technologies, when compared to the old system, without advanced technologies, is much better when the performance coefficient (CD) is higher than 1.30. Then, section S_1 is characterized by $CD > 1.30$. Analogously, all the established sections may be characterized as follows:

S_1: $CD > 1.30$ (the new system is **much better** than the old one);
S_2: $1.10 < CD < 1.30$ (the new system is **better** than the old one);
S_3: **$0.90 \leq CD \leq 1.10$** (the new system is **equivalent** to the old one);
S_4: $0.70 \leq CD < 0.90$ (the new system is **worse** than the old one);
S_5: $CD < 0.70$ (the new system is **much worse** than the old one).

It is important to observe that certain factors are not measurable. As examples of these factors, the strategic and the qualitative attributes may be mentioned. These factors (or indicators) cannot be translated by a value I and, therefore, for them, it is not possible to define the performance coefficient. In this case, the framing of the factor in the section will be done by an expert in the subject, who will say, for each considered factor, if the new system is better or worse than the old one. Then, the specialist, utilizing qualitative data, his/her knowledge, experience, intuition, common sense, etc., frames the factors in the sections, which are characterized analogously.

S_1: the new system is **much better** than the old one;
S_2: the new system is **better** than the old one;
S_3: the new system is **equivalent** to the old one;
S_4: the new system is **worse** than the old one;
S_5: the new system is **much worse** than the old one.

6.4.5 Construction of the Database

The database is constituted by the weights of the factors and by the degrees of the favorable evidence and contrary evidence attributed by the specialists to all the factors, in each one of the five established sections. Therefore, the database is constituted by the matrices $M_p = [P_i]$ and $M_A = [\mu_{i,j,k}] = [(a_{i,j,k}, b_{i,j,k})]$, of the weights and of the annotations.

It will be admitted that the matrix M_A, which constitutes one of the matrices of the database, is the same for the feasibility study for any new manufacturing system with advanced technologies (*CAD/CAM*—Computer-Automated Design and Manufacture; *GT/CM*—Group Technology and Cellular Manufacture; *RE*—Robotics Equipment; *FMS*—Flexible Manufacturing Systems; *AA*—Automated Assembly; *CIM*—Computer-Integrated Manufacture, etc.).

This is a simplifying hypothesis, which is being adopted in this work. And it is being adopted because we are also assuming that only the differentiation of the matrix of the weights is sufficient to have feasibility analysis of new manufacturing systems with sufficient validity and reliability. But there is nothing that prevents the KE from performing, for each analysis, a different matrix of annotations.

Therefore, in the present work, the matrix of the annotations, M_A, will always be the same, but the matrix of the weights, M_p, will vary at each new system with advanced technologies being analyzed, so that the specificity of each one is considered.

Table 6.16 constitutes matrix M_A, that is, it contains the degrees of the favorable evidence (or belief) and contrary evidence (or disbelief) attributed by the specialists to the factors, in the conditions established by the five sections. And the matrix of the annotations of the database. (Here, in the text, only one fragment of matrix M_A is presented; the complete matrix is found in Appendix C: BD-Sect. 6.4.)

The matrix of the weights for each chosen factor is also part of the database. The specialists attribute them, taking into account the new manufacturing system with advanced technologies which is being analyzed.

For the attribution of these weights, the KE may make some restrictions, such as: the weights must be integers of the interval [0; 10] (weight 0 means that the respective factor does not have any importance in the analysis being conducted and that, therefore, it must be excluded from this analysis).

A factor F_n may have greater importance in the feasibility analysis of a new system X than in another system Y. Then, this factor F_n must have greater weight in the first analysis than in the second one. With another factor F_m, exactly the contrary may occur: its importance may be smaller in the analysis of X than of Y. In this case, F_m must have smaller weight in the first analysis than in the second one. Therefore, the weight of a factor does not depend only on itself; it also depends on the system being analyzed. There may be a negligible factor (weight = 0) in the analysis of a system, but considerable (weight $\neq$ 0) in the analysis of another one. For the feasibility analysis of implementation of each system, a new attribution of weights to the chosen factors must be conducted, besides a new selection of the most important factors. Therefore, although we are assuming that the matrix of the annotations, $M_A = [\mu_{i,j,k}] = [(a_{i,j,k}, b_{i,j,k})]$, is the same, the matrix of the weights, $M_p = [P_i]$, must vary for the analysis of each new manufacturing system utilizing new technologies.

These weights are tremendously important. They indicate the relative importance of all the quantifiable and non quantifiable measures that constitute the company's success in the marketplace. These weights must be derived from a consensus of managerial opinions: technical, financial, marketing and administrative [22].

Table 6.16 Matrix of the annotations: degrees of the favorable evidence and contrary evidence attributed by the specialists to the factors, in each one of the sections

Factor F_i	Section S_j	Spec 1 $a_{i,1}$	$b_{i,1}$	Spec 2 $a_{i,2}$	$b_{i,2}$	Spec 3 $a_{i,3}$	$b_{i,3}$	Spec 4 $a_{i,4}$	$b_{i,4}$
F01	S1	1.00	0.00	0.90	0.10	1.00	0.10	0.90	0.00
	S2	0.70	0.20	0.80	0.30	0.80	0.20	0.70	0.30
	S3	0.50	0.50	0.60	0.50	0.60	0.40	0.50	0.40
	S4	0.30	0.70	0.40	0.60	0.40	0.70	0.30	0.60
	S5	0.00	1.00	0.10	0.80	0.20	0.90	0.20	1.00
F02	S1	1.00	0.05	0.95	0.15	1.00	0.10	0.85	0.00
	S2	0.75	0.25	0.85	0.25	0.85	0.30	0.73	0.35
	S3	0.55	0.45	0.55	0.45	0.65	0.40	0.45	0.55
	S4	0.35	0.65	0.4	0.65	0.35	0.75	0.24	0.65
	S5	0.00	0.95	0.15	0.75	0.15	0.85	0.25	1.00
F03	S1	0.92	0.08	0.98	0.18	0.88	0.12	0.82	0.07
	S2	0.67	0.23	0.83	0.27	0.77	0.18	0.63	0.28
	S3	0.52	0.47	0.57	0.48	0.62	0.43	0.52	0.45
	S4	0.27	0.73	0.45	0.65	0.37	0.67	0.33	0.57
	S5	0.05	0.98	0.17	0.83	0.18	0.02	0.21	0.95
F04	S1	0.95	0.11	1.00	0.21	0.91	0.15	0.85	0.10
	S2	0.70	0.26	0.86	0.30	0.80	0.21	0.66	0.31
	S3	0.55	0.50	0.60	0.51	0.65	0.46	0.55	0.48
	S4	0.30	0.76	0.48	0.68	0.40	0.70	0.36	0.60
	S5	0.08	1.01	0.20	0.86	0.21	0.05	0.24	0.98
F05	S1	0.88	0.04	0.94	0.14	0.84	0.08	0.78	0.03
	S2	0.63	0.19	0.79	0.23	0.73	0.14	0.59	0.24
	S3	0.48	0.43	0.53	0.44	0.58	0.39	0.48	0.41
	S4	0.23	0.69	0.41	0.61	0.33	0.63	0.29	0.53
	S5	0.01	0.94	0.13	0.79	0.14	0.00	0.17	0.91

Factor F_i	Section S_j	Spec 1 $a_{i,1}$	$b_{i,1}$	Spec 2 $a_{i,2}$	$b_{i,2}$	Spec 3 $a_{i,3}$	$b_{i,3}$	Spec 4 $a_{i,4}$	$b_{i,4}$
F17	S1	0.99	0.06	0.94	0.16	0.99	0.11	0.84	0.01
	S2	0.57	0.48	0.62	0.43	0.52	0.45	0.52	0.47
	S3	0.55	0.45	0.65	0.40	0.45	0.55	0.55	0.45
	S4	0.13	0.81	0.14	0.92	0.17	0.93	0.01	0.96
	S5	0.02	0.94	0.14	0.88	0.15	1.00	0.18	0.91
F18	S1	0.88	0.22	0.98	0.21	0.88	0.12	0.98	0.12
	S2	0.55	0.45	0.65	0.40	0.45	0.55	0.55	0.45
	S3	0.57	0.48	0.62	0.43	0.52	0.45	0.52	0.47
	S4	0.10	0.86	0.15	0.93	0.24	0.98	0.08	1.00
	S5	0.13	0.81	0.14	0.92	0.17	0.93	0.01	0.96
F19	S1	0.98	0.90	0.04	0.12	0.93	0.87	0.02	0.02
	S2	0.88	0.22	0.98	0.21	0.88	0.12	0.98	0.12
	S3	0.55	0.50	0.60	0.51	0.65	0.46	0.55	0.48
	S4	0.05	0.98	0.17	0.83	0.18	0.02	0.21	0.95
	S5	0.01	0.94	0.13	0.88	0.14	1.00	0.17	0.91
F20	S1	0.92	0.16	0.98	0.16	0.92	0.06	0.98	0.06
	S2	0.47	0.43	0.52	0.44	0.57	0.39	0.47	0.41
	S3	0.54	0.45	0.54	0.45	0.64	0.40	0.44	0.55
	S4	0.52	0.47	0.57	0.48	0.62	0.43	0.52	0.45
	S5	0.01	0.94	0.13	0.88	0.14	1.00	0.17	0.91
F21	S1	0.95	0.16	0.99	0.11	0.84	0.01	0.99	0.06
	S2	0.54	0.45	0.54	0.45	0.64	0.40	0.44	0.55
	S3	0.52	0.47	0.57	0.48	0.62	0.43	0.52	0.45
	S4	0.08	1.00	0.20	0.86	0.21	0.05	0.24	0.98
	S5	0.00	1.00	0.10	0.80	0.90	0.08	1.00	0.15

(continued)

Table 6.16 (continued)

Factor F_i	Section S_j	Spec 1 $a_{i,1}$	$b_{i,1}$	Spec 2 $a_{i,2}$	$b_{i,2}$	Spec 3 $a_{i,3}$	$b_{i,3}$	Spec 4 $a_{i,4}$	$b_{i,4}$
F06	S1	0.90	0.10	1.00	0.10	0.90	0.00	1.00	0.00
	S2	0.80	0.30	0.80	0.20	0.70	0.30	0.70	0.20
	S3	0.60	0.50	0.60	0.40	0.50	0.40	0.50	0.50
	S4	0.40	0.60	0.40	0.70	0.30	0.60	0.30	0.70
	S5	0.10	0.80	0.20	0.90	0.20	1.00	0.00	1.00
F07	S1	0.95	0.15	1.00	0.10	0.85	0.00	1.00	0.05
	S2	0.85	0.25	0.85	0.30	0.73	0.35	0.75	0.25
	S3	0.55	0.45	0.65	0.40	0.45	0.55	0.55	0.45
	S4	0.40	0.65	0.35	0.75	0.24	0.65	0.35	0.65
	S5	0.15	0.75	0.15	0.85	0.25	1.00	0.00	0.95
F08	S1	0.98	0.18	0.88	0.12	0.82	0.07	0.92	0.08
	S2	0.83	0.27	0.77	0.18	0.63	0.28	0.67	0.23
	S3	0.57	0.48	0.62	0.43	0.52	0.45	0.52	0.47
	S4	0.45	0.65	0.37	0.67	0.33	0.57	0.27	0.73
	S5	0.17	0.83	0.18	0.02	0.21	0.95	0.05	0.98
F09	S1	1.00	0.21	0.91	0.15	0.85	0.10	0.95	0.11
	S2	0.86	0.30	0.80	0.21	0.66	0.31	0.70	0.26
	S3	0.60	0.51	0.65	0.46	0.55	0.48	0.55	0.50
	S4	0.48	0.68	0.40	0.70	0.36	0.60	0.30	0.76
	S5	0.20	0.86	0.21	0.05	0.24	0.98	0.08	1.00
F10	S1	0.94	0.14	0.84	0.08	0.78	0.03	0.88	0.04
	S2	0.79	0.23	0.73	0.14	0.59	0.24	0.63	0.19
	S3	0.53	0.44	0.58	0.39	0.48	0.41	0.48	0.43
	S4	0.41	0.61	0.33	0.63	0.29	0.53	0.23	0.69
	S5	0.13	0.79	0.14	0.00	0.17	0.91	0.01	0.94

Factor F_i	Section S_j	Spec 1 $a_{i,1}$	$b_{i,1}$	Spec 2 $a_{i,2}$	$b_{i,2}$	Spec 3 $a_{i,3}$	$b_{i,3}$	Spec 4 $a_{i,4}$	$b_{i,4}$
F22	S1	1.00	0.21	0.91	0.15	0.85	0.10	0.95	0.11
	S2	0.54	0.45	0.54	0.45	0.64	0.40	0.44	0.55
	S3	0.52	0.47	0.57	0.48	0.62	0.43	0.52	0.45
	S4	0.47	0.43	0.52	0.44	0.57	0.39	0.47	0.41
	S5	0.00	1.00	0.10	0.80	0.90	0.08	1.00	0.15
F23	S1	0.97	0.90	0.03	0.12	0.92	0.87	0.01	0.02
	S2	0.55	0.45	0.55	0.45	0.65	0.40	0.45	0.55
	S3	0.48	0.43	0.53	0.44	0.58	0.39	0.48	0.41
	S4	0.08	0.83	0.18	0.95	0.21	0.95	0.05	0.98
	S5	0.06	0.86	0.11	0.93	0.20	0.98	0.08	1.00
F24	S1	0.90	0.10	0.95	0.10	0.90	0.00	0.95	0.00
	S2	0.88	0.04	0.94	0.14	0.84	0.08	0.78	0.03
	S3	0.53	0.44	0.58	0.39	0.48	0.41	0.48	0.43
	S4	0.08	0.83	0.18	0.95	0.21	0.95	0.05	0.98
	S5	0.06	0.86	0.11	0.93	0.20	0.98	0.08	1.00
F25	S1	0.98	0.90	0.04	0.12	0.93	0.87	0.02	0.02
	S2	0.88	0.04	0.94	0.14	0.84	0.08	0.78	0.03
	S3	0.53	0.44	0.58	0.39	0.48	0.41	0.48	0.43
	S4	0.08	0.83	0.18	0.95	0.21	0.95	0.05	0.98
	S5	0.06	0.86	0.11	0.93	0.20	0.98	0.08	1.00
F26	S1	0.92	0.16	0.98	0.16	0.92	0.06	0.98	0.06
	S2	0.55	0.50	0.60	0.51	0.65	0.46	0.55	0.48
	S3	0.53	0.44	0.58	0.39	0.48	0.41	0.48	0.43
	S4	0.14	0.86	0.19	0.93	0.28	0.98	0.12	1.00
	S5	0.06	0.86	0.11	0.93	0.20	0.98	0.08	1.00

(continued)

Table 6.16 (continued)

Factor	Section	Spec 1		Spec 2		Spec 3		Spec 4	
F_i	S_j	$a_{i,1}$	$b_{i,1}$	$a_{i,2}$	$b_{i,2}$	$a_{i,3}$	$b_{i,3}$	$a_{i,4}$	$b_{i,4}$
F11	S1	1.00	0.21	0.91	0.15	0.85	0.10	0.95	0.11
	S2	0.86	0.30	0.80	0.21	0.66	0.31	0.70	0.26
	S3	0.60	0.51	0.65	0.46	0.55	0.48	0.55	0.50
	S4	0.48	0.68	0.40	0.70	0.36	0.60	0.30	0.76
	S5	0.20	0.86	0.21	0.05	0.24	0.98	0.08	1.00
F12	S1	0.94	0.14	0.84	0.08	0.78	0.03	0.88	0.04
	S2	0.79	0.23	0.73	0.14	0.59	0.24	0.63	0.19
	S3	0.53	0.44	0.58	0.39	0.48	0.41	0.48	0.43
	S4	0.41	0.61	0.33	0.63	0.29	0.53	0.23	0.69
	S5	0.13	0.79	0.14	0.00	0.17	0.91	0.01	0.94
F13	S1	1.00	0.10	0.9	0.00	1.00	0.00	0.90	0.10
	S2	0.80	0.20	0.70	0.30	0.70	0.20	0.80	0.30
	S3	0.60	0.40	0.50	0.40	0.50	0.50	0.60	0.50
	S4	0.40	0.70	0.30	0.60	0.30	0.70	0.40	0.60
	S5	0.20	0.90	0.20	1.00	0.00	1.00	0.10	0.00
F14	S1	1.00	0.10	0.85	0.00	1.00	0.05	0.95	0.15
	S2	0.85	0.30	0.73	0.35	0.75	0.25	0.85	0.25
	S3	0.65	0.40	0.45	0.55	0.55	0.45	0.55	0.45
	S4	0.35	0.75	0.24	0.65	0.35	0.65	0.40	0.65
	S5	0.15	0.85	0.25	1.00	0.00	0.95	0.15	0.75
F15	S1	0.88	0.04	0.94	0.14	0.84	0.08	0.78	0.03
	S2	0.65	0.48	0.65	0.38	0.55	0.38	0.55	0.48
	S3	0.55	0.45	0.55	0.45	0.65	0.40	0.45	0.55
	S4	0.14	0.86	0.19	0.93	0.28	0.98	0.12	1.00
	S5	0.06	0.86	0.11	0.93	0.20	0.98	0.08	1.00

Factor	Section	Spec 1		Spec 2		Spec 3		Spec 4	
F_i	S_j	$a_{i,1}$	$b_{i,1}$	$a_{i,2}$	$b_{i,2}$	$a_{i,3}$	$b_{i,3}$	$a_{i,4}$	$b_{i,4}$
F27	S1	0.99	0.06	0.94	0.16	0.99	0.11	0.84	0.01
	S2	0.55	0.50	0.60	0.51	0.65	0.46	0.55	0.48
	S3	0.14	0.86	0.19	0.93	0.28	0.98	0.12	1.00
	S4	0.05	0.98	0.17	0.83	0.18	0.02	0.21	0.95
	S5	0.00	1.00	0.10	0.80	0.90	0.08	1.00	0.15
F28	S1	0.99	0.06	0.94	0.16	0.99	0.11	0.84	0.01
	S2	0.93	0.17	0.98	0.12	0.83	0.02	0.98	0.07
	S3	0.14	0.86	0.19	0.93	0.28	0.98	0.12	1.00
	S4	0.13	0.78	0.14	0.89	0.17	0.90	0.01	0.93
	S5	0.08	0.83	0.18	0.95	0.21	0.95	0.05	0.98
F29	S1	0.97	0.9	0.03	0.12	0.92	0.87	0.01	0.02
	S2	0.57	0.43	0.67	0.38	0.47	0.53	0.57	0.43
	S3	0.52	0.47	0.57	0.48	0.62	0.43	0.52	0.45
	S4	0.08	1.00	0.20	0.86	0.21	0.05	0.24	0.98
	S5	0.01	0.94	0.13	0.88	0.14	1.00	0.17	0.91
F30	S1	0.90	0.10	0.95	0.10	0.90	0.00	0.95	0.00
	S2	0.57	0.43	0.67	0.38	0.47	0.53	0.57	0.43
	S3	0.57	0.44	0.62	0.39	0.52	0.41	0.52	0.43
	S4	0.10	0.86	0.15	0.93	0.24	0.98	0.08	1.00
	S5	0.13	0.81	0.14	0.92	0.17	0.93	0.01	0.96
F31	S1	0.98	0.91	0.02	0.13	0.91	0.88	0.00	0.03
	S2	0.88	0.22	0.98	0.21	0.88	0.12	0.98	0.12
	S3	0.54	0.45	0.54	0.45	0.64	0.40	0.44	0.55
	S4	0.14	0.86	0.19	0.93	0.28	0.98	0.12	1.00
	S5	0.00	1.00	0.10	0.80	0.90	0.08	1.00	0.15

(continued)

Table 6.16 (continued)

Factor F_i	Section S_j	Spec 1 $a_{i,1}$	Spec 1 $b_{i,1}$	Spec 2 $a_{i,2}$	Spec 2 $b_{i,2}$	Spec 3 $a_{i,3}$	Spec 3 $b_{i,3}$	Spec 4 $a_{i,4}$	Spec 4 $b_{i,4}$
F16	S1	0.99	0.06	0.94	0.16	0.99	0.11	0.84	0.01
	S2	0.57	0.43	0.67	0.38	0.47	0.53	0.57	0.43
	S3	0.57	0.44	0.62	0.39	0.52	0.41	0.52	0.43
	S4	0.14	0.86	0.19	0.93	0.28	0.98	0.12	1.00
	S5	0.13	0.78	0.14	0.89	0.17	0.90	0.01	0.93

Factor F_i	Section S_j	Spec 1 $a_{i,1}$	Spec 1 $b_{i,1}$	Spec 2 $a_{i,2}$	Spec 2 $b_{i,2}$	Spec 3 $a_{i,3}$	Spec 3 $b_{i,3}$	Spec 4 $a_{i,4}$	Spec 4 $b_{i,4}$
F32	S1	0.99	0.25	0.90	0.19	0.84	0.14	0.94	0.15
	S2	0.55	0.45	0.65	0.40	0.45	0.55	0.55	0.45
	S3	0.57	0.48	0.62	0.43	0.52	0.45	0.52	0.47
	S4	0.14	0.86	0.19	0.93	0.28	0.98	0.12	1.00
	S5	0.06	0.86	0.11	0.93	0.20	0.98	0.08	1.00

For that reason, the weights are attributed by the specialists, in consensus among them or considering the average of the weights each one attributes, isolatedly (see Sect. 4.2.4).

It is convenient that the specialists called for the constitution of the database have different and complementary majors so that the different aspects of the problem are taken into account. For example, in the present work, a set of four specialists will be considered, constituted as follows: E_1—a production engineer (technician); E_2—a marketing professional; E_3—a finance professional; and E_4—an industrial administrator.

6.4.6   Feasibility Analysis of the Implementation of a Flexible Manufacturing System

To demonstrate the feasibility analysis methodology for decision making, based on the logic Eτ, it will be applied in the feasibility study of the implementation of a Flexible Manufacturing System (FMS).

It must be remembered that "FMS are groups of production machines organized in sequence, connected by material handling and transference automated machines, and integrated by a computer system." [57].

For this case, based on the chapter Justifying Capital Investment of [22], the influence factors listed below were chosen as fundamental.

It will be assumed that, in consensus, the specialists attributed to each one of these factors, according to their importance in the decision about the feasibility of implementation of the FMS, the weights placed in brackets to the left side of each one of them, on a scale from 1 to 10 (the factors with weight 0 (zero) in the study for implementation of the FMS" are already excluded from this list). This assumption was also based on the mentioned chapter. Therefore, the numbers in brackets on the left side of each factor constitute the matrix of the weights, $M_p = [P_i]$, which completes the database, for this case.

(5) F_{01}—technological reputation
(4) F_{02}—market share occupied by the company
(6) F_{03}—competitive position of the company inside the market occupied by it
(4) F_{04}—innovation of the product produced by the company
(10) F_{07}—net present value (NPV)
(5) F_{08}—payback period
(3) F_{11}—product heterogeneity
(3) F_{12}—product reliability
(1) F_{18}—additional indirect labor costs

(1) F_{19}—material cost
(2) F_{20}—capital investment cost
(1) F_{24}—answer time
(2) F_{25}—preparation time
(3) F_{29}—number of produced pieces

Researches conducted by specialists and using surveys at the companies that have already adopted FMS systems and have them up and running enable us to verify in which section, S_{pj}, each one of these factors is (see column 3 of Table 6.17). These sections constitute the researched research, $M_{pq} = [S_{pj}]$.

The chosen specialists will be distributed in two groups: group A, constituted by experts E_1 (production engineer) and E_2 (marketing professional), and Group B, constituted by specialists E_3 (finance professional) and E_4 (industrial administrator). This way, the arrangement for application of the **MAX** (maximizing) and **MIN** (minimizing) operators is the following:

$$\mathbf{MIN}\{\mathbf{MAX}[(E_1), (E_2)], \mathbf{MAX}[(E_3), (E_4)]\} \text{ or } \mathbf{MIN}\{G_A, G_B\}$$

For the decision making, considering that the replacement of a traditional system for an FMS one involves a high investment with great risk of loss for the company, a high value will be adopted for the requirement level, equal to 0.75. That means that the decision will be made if $|H| \geq 0.75$, that is, the segments defined by the condition $|H| = 0.75$ are being adopted as limit lines of truth and falsity. Thus, the para-analyzing algorithm and the decision rule are already determined (see Sect. 6.4.2, Fig. 6.12).

After the framing of the factors in the sections by means of the field research has been done (column 3 of Table 6.17), the weights (column 2) and the criteria (decision rule) have been established, with the aid of the CP of the PDM, the experts' opinions (degrees of favorable evidence and contrary evidence) regarding the enterprise in the conditions of the factors, translated by the researched sections, are sought in the database (Table 6.16), obtaining columns 4–11 of Table 6.17. Having obtained these opinions, the same CP of the PDM applies the maximization (**MAX**) and minimization (**MIN**) techniques of the logic $E\tau$ to each one of the factors, obtaining the resultant degrees of favorable evidence and contrary evidence (columns 16 and 17), which enable us to calculate the degree of certainty for each factor (column 18). With this value, inside the established requirement level (0.75), the CP of the PDM itself concludes if the factor contributes for the feasibility or unfeasibility of the FMS system or if the factor is inconclusive (column 20).

6.4.7 Analysis of the Results

In this case, the feasibility analysis of implementation of the FMS, presented as an example, at the requirement level of 0.75, seven factors (F_{01}, F_{03}, F_{07}, F_{08}, F_{12}, F_{25} and F_{29}) were favorable to the enterprise, indicates the feasibility of its implementation; one, unfavorable (F_{19}), which indicates the unfeasibility of its implementation; and six are inconclusive (F_{02}, F_{04}, F_{11}, F_{18}, F_{20} and F_{24}), everything at the established requirement level (0.75).

The joint influence of all these factors on the decision of the feasibility of the enterprise may be summarized by the center of gravity (W) of the points that

Table 6.17 Calculations and result analysis done by CP of the PDM with level of requirement **0.75** (in this level the analysis showed **NOT CONCLUSIVE**)

1	2	3	4	5	6	7	8	9	10	11	12	13	14	15	16	17	18	19	20
			Group A				Group B				E1 MAX E2		E3 MAX E4		A MIN B		Level of requirement		0.75
											A		B				Conclusions		
F_i	P_i	S_j	E_1		E_2		E_3		E_4										Decision
			$a_{i,1}$	$b_{i,1}$	$a_{i,2}$	$b_{i,2}$	$a_{i,3}$	$b_{i,3}$	$a_{i,4}$	$b_{i,4}$	$a_{i,gA}$	$b_{i,gA}$	$a_{i,gB}$	$b_{i,gB}$	$a_{i,R}$	$b_{i,R}$	H_{cert}	G_{contr}	
F_{01}	5	S_1	1.00	0.00	0.90	0.10	1.00	0.10	0.90	0.00	1.00	0.00	1.00	0.00	1.00	0.00	1.00	0.00	VIABLE
F_{02}	4	S_2	0.75	0.25	0.85	0.25	0.85	0.30	0.73	0.35	0.85	0.25	0.85	0.30	0.85	0.30	0.55	0.15	NOT CONCLUSIVE
F_{03}	6	S_1	0.92	0.08	0.98	0.18	0.88	0.12	0.82	0.07	0.98	0.08	0.88	0.07	0.88	0.08	0.80	−0.04	VIABLE
F_{04}	4	S_2	0.70	0.26	0.86	0.30	0.80	0.21	0.66	0.31	0.86	0.26	0.80	0.21	0.80	0.26	0.54	0.06	NOT CONCLUSIVE
F_{07}	10	S_1	0.95	0.15	1.00	0.10	0.85	0.00	1.00	0.05	1.00	0.10	1.00	0.00	1.00	0.10	0.90	0.10	VIABLE
F_{08}	5	S_1	0.98	0.18	0.88	0.12	0.82	0.07	0.92	0.08	0.98	0.12	0.92	0.07	0.92	0.12	0.80	0.04	VIABLE
F_{11}	3	S_2	0.86	0.30	0.80	0.21	0.66	0.31	0.70	0.26	0.86	0.21	0.70	0.26	0.70	0.26	0.44	−0.04	NOT CONCLUSIVE
F_{12}	3	S_1	0.94	0.14	0.84	0.08	0.78	0.03	0.88	0.04	0.94	0.08	0.88	0.03	0.88	0.08	0.80	−0.04	VIABLE
F_{18}	1	S_3	0.57	0.48	0.62	0.43	0.52	0.45	0.52	0.47	0.62	0.43	0.52	0.45	0.52	0.45	0.07	−0.03	NOT CONCLUSIVE
F_{19}	1	S_5	0.01	0.94	0.13	0.88	0.14	1.00	0.17	0.91	0.13	0.88	0.17	0.91	0.13	0.91	−0.78	0.04	UNVIABLE
F_{20}	2	S_2	0.47	0.43	0.52	0.44	0.57	0.39	0.47	0.41	0.52	0.43	0.57	0.39	0.52	0.43	0.09	−0.05	NOT CONCLUSIVE
F_{24}	1	S_4	0.14	0.86	0.19	0.93	0.18	0.02	0.21	0.95	0.19	0.86	0.21	0.02	0.19	0.86	−0.67	0.05	NOT CONCLUSIVE
F_{25}	2	S_2	0.88	0.04	0.94	0.14	0.84	0.08	0.78	0.03	0.94	0.04	0.84	0.03	0.84	0.04	0.80	−0.12	VIABLE
F_{29}	3	S_1	0.97	0.90	0.03	0.12	0.92	0.87	0.01	0.02	0.97	0.12	0.92	0.02	0.92	0.12	0.80	0.04	VIABLE
50			Baricenter W: weighted averages of the resultant degrees													0.851	0.177	0.674	0.028	NOT CONCLUSIVE

represent them in the para-analyzer algorithm. Thus, to have the final and global conclusion of the analysis, taking into account the combined influence of all the factors, the CP of the PDM calculates the degrees of favorable evidence and contrary evidence of the center of gravity (W). They are obtained calculating the weighted averages of the resultant degrees of favorable evidence and contrary evidence for each one of the factors. With the degrees of favorable evidence (a_W) and of contrary evidence (b_W) of the center of gravity, (last line of columns 16 and 17), its degree of certainty (last line of column 18) is calculated, which enables the final decision (last line column 20): the analysis conducted to study the feasibility of the implementation of the FMS is inconclusive, at the established requirement level (0.75).

Some calculations performed by the CP of the PDM, in Table 6.17, may be highlighted. For the center of gravity, the degree of favorable evidence ($a_W = 0.851$) and the degree of contrary evidence ($b_W = 0.177$) were obtained. From these values, its degree of certainty was calculated: $H_W = a_W - b_W = 0.851 - 0.177 = 0.674$. Once $-0.75 < 0.674 < 0.75$, applying the decision rule, it is inferred that the analysis is inconclusive, that is, the analysis does not suggest the feasibility nor for the unfeasibility of the enterprise execution.

The analysis conducted by the application of the decision rule, for each factor separately or for the center of gravity (which considers the joint influence of all the factors), may also be performed through the para-analyzer algorithm. Just plot the resultant degrees of favorable evidence and contrary evidence (columns 16 and 17) in lattice τ, as seen in Fig. 6.13. In this figure, we notice that the representative

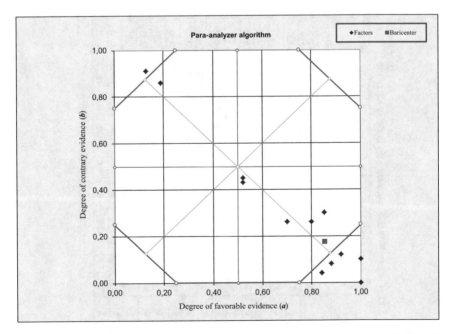

Fig. 6.13 Analysis of the results by the para-analyzing device at the requirement level of **0.75** (at this level, the analysis was **inconclusive**)

points of seven factors belong to the truth region (suggest the implementation of the FMS); one belongs to the falsity region (suggests the non-implementation of the FMS) and the other seven belong to different regions, being, therefore, inconclusive.

If the requirement level is altered, reducing it to 0.60, it is observed that the center of gravity now belongs to the truth region (Fig. 6.14) and the conducted analysis is now conclusive. We reach a favorable decision to the implementation of the FMS system, that is, the implementation of the FMS is feasible, at the requirement level 0.60.

In Fig. 6.14, we notice that, at the requirement level 0.60, seven factors belong to the truth region (in the figure only five diamonds appear, because there are two coincidences) and suggest the implementation of the FMS; two belong to the falsity region and suggest the non-implementation of the FMS, and the other five belong to different regions, being, therefore, inconclusive.

The calculation table of the CP of the PDM, in this case, is turned into Table 6.18, which differs from Table 6.17 only in the decision column (column 20), as only the requirement level was altered.

6.5   Diagnosis Prediction

In this paragraph, one more application for the Paraconsistent Decision Method (PDM) in the aid to decision making will be analyzed. It will be applied in diagnosis prediction, which is also decision making, diagnosing is nothing more than deciding, among the available options, which one of them is the most likely or the one with most evidence. The purpose of this other application is to show a variant of the PDM and that it may be applied even in analysis of problems that involve very large databases.

For the presentation of the process, we focused on the problem of the prediction of a medical diagnosis, although the method may be applied identically for the prediction of other diagnoses, such as defects in industrial machines, airplanes, ships, cars, trucks, etc. Therefore, in this application, the prediction of diagnoses of diseases will be performed based on the symptoms (or signals) presented by the patient. This prediction is critical to be utilized, for example, in the screening of large hospitals, facilitating the personnel who work there to conduct the patient's submission to the specialized sector in the foreseen disease. Evidently, this forecast is not intended to replace the diagnosis performed by a doctor or medical board, under any circumstances.

Table 6.18 Calculations and result analysis done by CP of the PDM with level of requirement **0.60** (in this level the analysis showed **VIABLE**)

1	2	3	4	5	6	7	8	9	10	11	12	13	14	15	16	17	18	19	20
F_i	P_i	S_j	Group A				Group B				A		B		A MIN B		Conclusions		
			E_1		E_2		E_3		E_4		E1 MAX E2		E3 MAX E4				Level of requirement		0.60
			$a_{i,1}$	$b_{i,1}$	$a_{i,2}$	$b_{i,2}$	$a_{i,3}$	$b_{i,3}$	$a_{i,4}$	$b_{i,4}$	$a_{i,gA}$	$b_{i,gA}$	$a_{i,gB}$	$b_{i,gB}$	$a_{i,LR}$	$b_{i,LR}$	H_{cert}	G_{contr}	Decision
F01	5	S1	1.00	0.00	0.90	0.10	1.00	0.10	0.90	0.00	1.00	0.00	1.00	0.00	1.00	0.00	1.00	0.00	VIABLE
F02	4	S2	0.75	0.25	0.85	0.25	0.85	0.30	0.73	0.35	0.85	0.25	0.85	0.30	0.85	0.30	0.55	0.15	NOT CONCLUSIVE
F03	6	S1	0.92	0.08	0.98	0.18	0.88	0.12	0.82	0.07	0.98	0.08	0.88	0.07	0.88	0.08	0.80	-0.04	VIABLE
F04	4	S2	0.70	0.26	0.86	0.30	0.80	0.21	0.66	0.31	0.86	0.26	0.80	0.21	0.80	0.26	0.54	0.06	NOT CONCLUSIVE
F07	10	S1	0.95	0.15	1.00	0.10	0.85	0.00	1.00	0.05	1.00	0.10	1.00	0.00	1.00	0.10	0.90	0.10	VIABLE
F08	5	S1	0.98	0.18	0.88	0.12	0.82	0.07	0.92	0.08	0.98	0.12	0.92	0.07	0.92	0.12	0.80	0.04	VIABLE
F11	3	S2	0.86	0.30	0.80	0.21	0.66	0.31	0.70	0.26	0.86	0.21	0.70	0.26	0.70	0.26	0.44	-0.04	NOT CONCLUSIVE
F12	3	S1	0.94	0.14	0.84	0.08	0.78	0.03	0.88	0.04	0.94	0.08	0.88	0.03	0.88	0.08	0.80	-0.04	NOT CONCLUSIVE
F18	1	S3	0.57	0.48	0.62	0.43	0.52	0.45	0.52	0.47	0.62	0.43	0.52	0.45	0.52	0.45	0.07	-0.03	NOT CONCLUSIVE
F19	1	S5	0.01	0.94	0.13	0.88	0.14	1.00	0.17	0.91	0.13	0.88	0.17	0.91	0.13	0.91	-0.78	0.04	UNVIABLE
F20	2	S2	0.47	0.43	0.52	0.44	0.57	0.39	0.47	0.41	0.52	0.43	0.57	0.39	0.52	0.43	0.09	-0.05	NOT CONCLUSIVE
F24	1	S4	0.14	0.86	0.19	0.93	0.18	0.02	0.21	0.95	0.19	0.86	0.21	0.02	0.19	0.86	-0.67	0.05	UNVIABLE
F25	2	S2	0.88	0.04	0.94	0.14	0.84	0.08	0.78	0.03	0.94	0.04	0.84	0.03	0.84	0.04	0.80	-0.12	VIABLE
F29	3	S1	0.97	0.90	0.03	0.12	0.92	0.87	0.01	0.02	0.97	0.12	0.92	0.02	0.92	0.12	0.80	0.04	VIABLE
50			Baricenter W: weighted averages of the resultant degrees													0.851	0.177	0.674	0.028	VIABLE

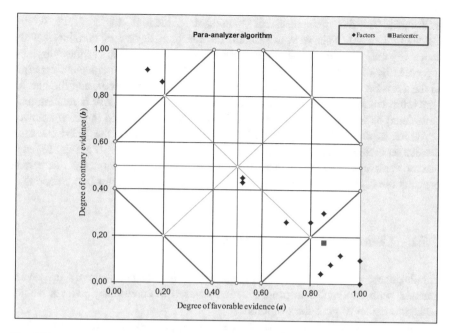

Fig. 6.14 Analysis of the results by the para-analyzing device at the requirement level of **0.60** (at this level, the analysis indicated that the enterprise is **feasible**)

For its purpose, a database formed by the opinions of medicine experts will be utilized, constituted by the values of favorable evidence (or degrees or belief) and by the values of contrary evidence (or degrees of disbelief) that each medical specialist attributes to each disease, when a certain symptom (or signal) is presented by the patient.

In this approach, for the presentation of the method, non-real data and a set of thirty-two possible diseases D_i ($1 \leq i \leq 32$) (ordered from AA to BF) will be considered, which will be related to another set of thirty symptoms S_j ($1 \leq j \leq 30$) (ordered from S_{01} to S_{30}).

Having the database (Table 6.19), the process consists, firstly, in performing the field research, verifying the symptoms the patient presents, using an interview conducted with him/her (anamnesis). Then, knowing the symptoms, the CP of the PDM applies the maximization (**MAX** operator) and minimization (**MIN** operator) techniques of the paraconsistent annotated evidential logic (Table 6.20—Calculation table). This way, we obtain the resultant degree of certainty (of the center of gravity) for each one of the thirty-two diseases as a result of the symptoms presented by the patient. The CP of the PDM "takes" all these degrees of certainty to the decision table (Table 6.21), compares them with each other, and the one for which the value of the degree of certainty of the center of gravity is maximum is considered as the diagnosis prediction. The process is presented in detail below.

Which is done, therefore, is nothing more than applying the PDM for each disease separately, obtaining the degree of certainty of the center of gravity for each one of them and comparing these obtained degrees of certainty; the one which presents the highest degree of certainty is regarded as the diagnosis prediction.

6.5.1 Construction of the Database

For the construction of the database, specialists in medicine, mainly in general practice, with experience in propaedeutic and used to conducting patients' anamnesis, are called to give their opinion.

Using their knowledge, experiences, sensitivity, intuition, common sense, etc., they must attribute values of favorable evidence (or degree of belief) and contrary evidence (or degree of disbelief) for each one of the thirty-two diseases for each one of the thirty symptoms, chosen to constitute the database.

In the case, as already said, for the presentation of the paragraph, thirty-two diseases and thirty symptoms were chosen. Four specialists were utilized (from E_1 to E_4), chosen by the KE. Therefore, the database will be constituted of $32 \times 30 \times 4 \times 2 = 7,680$ pieces of data, presented in a table with 960 lines and 8 columns.

Out of these 7,680 pieces of data, half represents values of favorable evidences and the other half, values of contrary evidences. A small part of the database is shown in Table 6.19, emphasizing its beginning with disease AA and its end with disease BF.

The complete database, with the thirty-two diseases and the thirty symptoms, is presented in Appendix C: BD-Sect. 6.5.

Table 6.19 Database: values of the degrees of the favorable evidence and contrary evidence attributed two the specialists for the thirty-two diseases, for each one of the thirty symptoms

Desease Symptom		E_1		E_2		E_3		E_4	
D_i	S_j	$a_{i,j,1}$	$b_{i,j,1}$	$a_{i,j,2}$	$b_{i,j,2}$	$a_{i,j,3}$	$b_{i,j,3}$	$a_{i,j,4}$	$b_{i,j,4}$
AA	S_{01}	0.88	0.04	0.94	0.14	0.84	0.08	0.78	0.03
AA	S_{02}	1.00	0.04	0.95	0.15	1.00	0.10	0.85	0.00
AA	S_{03}	0.90	0.10	0.96	0.20	0.86	0.14	0.80	0.09
AA	S_{04}	0.97	0.14	1.00	0.24	0.93	0.19	0.87	0.13
AA	S_{05}	0.98	0.91	0.02	0.13	0.91	0.88	0.00	0.03
AA	S_{06}	0.65	0.48	0.65	0.38	0.55	0.38	0.55	0.48
AA	S_{07}	0.57	0.43	0.67	0.38	0.47	0.53	0.57	0.43
AA	S_{08}	0.57	0.44	0.62	0.39	0.52	0.41	0.52	0.43
AA	S_{09}	0.14	0.86	0.19	0.93	0.28	0.98	0.12	1.00
AA	S_{10}	0.13	0.78	0.14	0.89	0.17	0.90	0.01	0.93
AA	S_{11}	0.94	0.14	0.84	0.08	0.78	0.03	0.88	0.04
AA	S_{12}	0.95	0.15	1.00	0.10	0.85	0.00	1.00	0.04
AA	S_{13}	0.96	0.20	0.86	0.14	0.80	0.09	0.90	0.10
AA	S_{14}	1.00	0.24	0.93	0.19	0.87	0.13	0.97	0.14
AA	S_{15}	0.02	0.13	0.91	0.88	0.00	0.03	0.98	0.91
AA	S_{16}	0.65	0.38	0.55	0.38	0.55	0.48	0.65	0.48
AA	S_{17}	0.67	0.38	0.47	0.53	0.57	0.43	0.57	0.43
AA	S_{18}	0.62	0.39	0.52	0.41	0.52	0.43	0.57	0.44
AA	S_{19}	0.19	0.93	0.28	0.98	0.12	1.00	0.14	0.86
AA	S_{20}	0.14	0.89	0.17	0.90	0.01	0.93	0.13	0.78
AA	S_{21}	0.84	0.08	0.78	0.03	0.88	0.04	0.94	0.14
AA	S_{22}	1.00	0.10	0.85	0.00	1.00	0.04	0.95	0.15
AA	S_{23}	0.86	0.14	0.80	0.09	0.90	0.10	0.96	0.20
AA	S_{24}	0.93	0.19	0.87	0.13	0.97	0.14	1.00	0.24
AA	S_{25}	0.91	0.88	0.00	0.03	0.98	0.91	0.02	0.13
AA	S_{26}	0.55	0.38	0.55	0.48	0.65	0.48	0.65	0.38
AA	S_{27}	0.47	0.53	0.57	0.43	0.57	0.43	0.67	0.38
AA	S_{28}	0.52	0.41	0.52	0.43	0.57	0.44	0.62	0.39
AA	S_{29}	0.28	0.98	0.12	1.00	0.14	0.86	0.19	0.93
AA	S_{30}	0.17	0.9	0.01	0.93	0.13	0.78	0.14	0.89
AA	S_{99}	0.00	0.00	0.00	0.00	0.00	0.00	0.00	0.00
AB	S_{01}	0.02	0.94	0.14	0.88	0.15	1.00	0.18	0.91
AB	S_{02}	0.99	0.06	0.94	0.16	0.99	0.11	0.84	0.01
AB	S_{03}	0.91	0.09	0.97	0.19	0.97	0.13	0.81	0.08
AB	S_{04}	0.93	0.15	0.96	0.25	...	...	...	...

... ...

(continued)

Table 6.19 (continued)

Desease	Symptom	E_1		E_2		E_3		E_4	
D_i	S_j	$a_{i,j,1}$	$b_{i,j,1}$	$a_{i,j,2}$	$b_{i,j,2}$	$a_{i,j,3}$	$b_{i,j,3}$	$a_{i,j,4}$	$b_{i,j,4}$
BF	S_{01}	0.67	0.38	0.47	0.53	0.57	0.43	0.57	0.43
BF	S_{02}	0.62	0.39	0.52	0.41	0.52	0.43	0.57	0.44
BF	S_{03}	0.19	0.93	0.28	0.98	0.12	1.00	0.14	0.86
BF	S_{04}	0.14	0.92	0.17	0.93	0.01	0.96	0.13	0.81
BF	S_{05}	0.15	1.00	0.18	0.91	0.02	0.94	0.14	0.88
BF	S_{06}	0.99	0.11	0.84	0.01	0.99	0.06	0.94	0.16
BF	S_{07}	0.18	0.02	0.21	0.95	0.05	0.98	0.17	0.83
BF	S_{08}	0.65	0.46	0.55	0.48	0.55	0.50	0.60	0.51
BF	S_{09}	0.93	0.87	0.02	0.02	0.98	0.90	0.04	0.12
BF	S_{10}	0.89	0.06	0.99	0.06	0.89	0.15	0.99	0.15
BF	S_{11}	0.84	0.01	0.99	0.06	0.95	0.16	0.99	0.11
BF	S_{12}	0.81	0.07	0.91	0.08	0.97	0.18	0.87	0.12
BF	S_{13}	0.20	0.98	0.08	1.00	0.06	0.86	0.11	0.93
BF	S_{14}	0.48	0.41	0.48	0.43	0.53	0.44	0.58	0.39
BF	S_{15}	0.48	0.43	0.53	0.44	0.58	0.39	0.48	0.41
BF	S_{16}	1.00	0.04	0.95	0.15	1.00	0.10	0.85	0.00
BF	S_{17}	0.91	0.09	0.97	0.19	0.97	0.13	0.81	0.08
BF	S_{18}	0.93	0.15	0.96	0.25	0.87	0.19	0.81	0.14
BF	S_{19}	0.00	1.00	0.10	0.80	0.90	0.08	1.00	0.15
BF	S_{20}	0.89	0.15	0.99	0.15	0.89	0.06	0.99	0.06
BF	S_{21}	0.93	0.17	0.98	0.12	0.83	0.02	0.98	0.07
BF	S_{22}	0.57	0.44	0.62	0.39	0.52	0.41	0.52	0.43
BF	S_{23}	0.10	0.86	0.15	0.93	0.24	0.98	0.08	1.00
BF	S_{24}	0.13	0.81	0.14	0.92	0.17	0.93	0.01	0.96
BF	S_{25}	0.52	0.44	0.57	0.39	0.47	0.41	0.47	0.43
BF	S_{26}	0.54	0.45	0.64	0.40	0.44	0.55	0.54	0.45
BF	S_{27}	0.98	0.18	0.88	0.12	0.82	0.07	0.92	0.08
BF	S_{28}	1.00	0.24	0.93	0.19	0.87	0.13	0.97	0.14
BF	S_{29}	0.04	0.11	0.93	0.86	0.02	0.01	0.98	0.89
BF	S_{30}	0.55	0.38	0.55	0.48	0.65	0.48	0.65	0.38

6.5.2   Calculation of the Resultant Degree of Certainty for Each Disease as a Result of the Symptoms Presented by the Patient

For the application of the paraconsistent annotated logic Eτ techniques, the four chosen specialists must be distributed in groups, according to their characteristics. Thus, for example, if one of the specialists is highly renowned and reputable, he/she may constitute one group alone; if two specialists have approximately the same major and the same knowledge and experience level, they may constitute one group, etc. The experts are distributed in two groups: Group A, constituted by specialists E_1 and E_2, and Group B, constituted by experts E_3 and E_4.

The maximization (**MAX** operator) is applied intragroup, that is, inside group A and group B, and the minimization (**MIN** operator) is applied intergroup, that is, among the results obtained by the maximization applied to groups A and B. Thus, the scheme being adopted to implement the techniques of logic Eτ is the following:

$$\textbf{MIN}\{\textbf{MAX}[(E_1,),(E_2)],\textbf{MAX}[(E_3),(E_4)]\} \text{ or } \textbf{MIN}\{G_A,G_B\}$$

The calculation of the resultant degree of certainty for each disease, considering symptoms presented by the patient, is performed with the aid of the CP of the PDM. Once the symptoms presented by the patient (S_{pj}) are known, they are placed in column 2 of the decision Table 6.20. From that moment on, the CP of the PDM (*i*) "transports" the values of this column to the corresponding column of the calculation tables (column 2 of Table 6.21); (*ii*) searches for the specialists' opinions in the database (Table 6.19), bringing them to the calculation table (columns 3–10 of Table 6.21); (*iii*) applies the techniques of logic Eτ and performed the calculations (columns 11–18 of Table 6.21), obtaining the degree of certainty of each disease as a result of each symptom, isolatedly (column 17 of Table 6.21); then, the CP of the PDM (*iv*) "takes" these results referring to the centers of gravity to the decision table (column 4 of Table 6.20) and (*v*) displays the disease with the highest degree of certainty (column 5). It is the diagnosis prediction.

As we may observe, the CP of the PDM performs practically everything, since the data search until the final decision making. Therefore, the only task to be performed is to feed it, verifying which are the symptoms presented by the patient and placing them in column 2 of the decision Table 6.20.

With that, the resultant degrees of favorable evidence and contrary evidence for each disease are obtained, about each symptom presented by the patient (columns 15 and 16 of Table 6.21). These resultant values, when plotted in the para-analyzing algorithm, result in points, each one of them representing the influence of a symptom presented by the patient in the considered disease. The center of gravity of these points translates the combined effect of all the symptoms submitted by the patient in the considered disease.

For the continuity of the exposition of this section, we will consider a patient that presents the **twelve** symptoms below: S_{01}, S_{02}, S_{03}, S_{05}, S_{07}, S_{09}, S_{12}, S_{15}, S_{18}, S_{22},

Table 6.20 Calculation table for AA (D₁) desease

1	2	3	4	5	6	7	8	9	10	11	12	13	14	15	16	17	18
Di	Sj	Group A				Group B				Group A		Group B		Resultant		L. of requir.	
		E1		E2		E3		E4		MAX [E1, E2]		MAX [E3, E4]		MIN {GA, GB}		0.60	
		$a_{i,j,1}$	$b_{i,j,1}$	$a_{i,j,2}$	$b_{i,j,2}$	$a_{i,j,3}$	$b_{i,j,3}$	$a_{i,j,4}$	$b_{i,j,4}$	$a_{i,j,A}$	$b_{i,j,A}$	$a_{i,j,B}$	$b_{i,j,B}$	$a_{i,j,R}$	$b_{i,j,R}$	H	G
AA	S01	0.88	0.04	0.94	0.14	0.84	0.08	0.78	0.03	0.94	0.04	0.84	0.03	0.84	0.04	0.80	−0.12
AA	S02	1.00	0.04	0.95	0.15	1.00	0.10	0.85	0.00	1.00	0.04	1.00	0.00	1.00	0.04	0.96	0.04
AA	S03	0.90	0.10	0.96	0.20	0.86	0.14	0.8	0.09	0.96	0.10	0.86	0.09	0.86	0.10	0.76	−0.04
AA	S05	0.98	0.91	0.02	0.13	0.91	0.88	0.00	0.03	0.98	0.13	0.91	0.03	0.91	0.13	0.78	0.04
AA	S07	0.57	0.43	0.67	0.38	0.47	0.53	0.57	0.43	0.67	0.38	0.57	0.43	0.57	0.43	0.14	0.00
AA	S09	0.14	0.86	0.19	0.93	0.28	0.98	0.12	1.00	0.19	0.86	0.28	0.98	0.19	0.98	−0.79	0.17
AA	S12	0.95	0.15	1.00	0.10	0.85	0.00	1.00	0.04	1.00	0.10	1.00	0.00	1.00	0.10	0.90	0.10
AA	S15	0.02	0.13	0.91	0.88	0.00	0.03	0.98	0.91	0.91	0.13	0.98	0.03	0.91	0.13	0.78	0.04
AA	S18	0.62	0.39	0.52	0.41	0.52	0.43	0.57	0.44	0.62	0.39	0.57	0.43	0.57	0.43	0.14	0.00
AA	S22	1.00	0.10	0.85	0.00	1.00	0.04	0.95	0.15	1.00	0.00	1.00	0.04	1.00	0.04	0.96	0.04
AA	S26	0.55	0.38	0.55	0.48	0.65	0.48	0.65	0.38	0.55	0.38	0.65	0.38	0.55	0.38	0.17	−0.07
AA	S30	0.17	0.90	0.01	0.93	0.13	0.78	0.14	0.89	0.17	0.90	0.14	0.78	0.14	0.90	−0.76	0.04
AA	S99	0.00	0.00	0.00	0.00	0.00	0.00	0.00	0.00	0.00	0.00	0.00	0.00	0.00	0.00	0.00	−1.00
AA	S99	0.00	0.00	0.00	0.00	0.00	0.00	0.00	0.00	0.00	0.00	0.00	0.00	0.00	0.00	0.00	−1.00
AA	S99	0.00	0.00	0.00	0.00	0.00	0.00	0.00	0.00	0.00	0.00	0.00	0.00	0.00	0.00	0.00	−1.00
AA	S99	0.00	0.00	0.00	0.00	0.00	0.00	0.00	0.00	0.00	0.00	0.00	0.00	0.00	0.00	0.00	−1.00
AA	S99	0.00	0.00	0.00	0.00	0.00	0.00	0.00	0.00	0.00	0.00	0.00	0.00	0.00	0.00	0.00	−1.00
AA	S99	0.00	0.00	0.00	0.00	0.00	0.00	0.00	0.00	0.00	0.00	0.00	0.00	0.00	0.00	0.00	−1.00

(continued)

Table 6.20 (continued)

1	2	3	4	5	6	7	8	9	10	11	12	13	14	15	16	17	18
Di	Sj	Group A		E2		Group B		E4		Group A		Group B		Resultant		L. of requir.	
		E1				E3				MAX [E1, E2]		MAX [E3, E4]		MIN {GA, GB}		0.60	
		$a_{ij,1}$	$b_{ij,1}$	$a_{ij,2}$	$b_{ij,2}$	$a_{ij,3}$	$b_{ij,3}$	$a_{ij,4}$	$b_{ij,4}$	$a_{ij,A}$	$b_{ij,A}$	$a_{ij,B}$	$b_{ij,B}$	$a_{ij,R}$	$b_{ij,R}$	H	G
AA	S99	0.00	0.00	0.00	0.00	0.00	0.00	0.00	0.00	0.00	0.00	0.00	0.00	0.00	0.00	0.00	-1.00
AA	S99	0.00	0.00	0.00	0.00	0.00	0.00	0.00	0.00	0.00	0.00	0.00	0.00	0.00	0.00	0.00	-1.00
AA	S99	0.00	0.00	0.00	0.00	0.00	0.00	0.00	0.00	0.00	0.00	0.00	0.00	0.00	0.00	0.00	-1.00
AA	S99	0.00	0.00	0.00	0.00	0.00	0.00	0.00	0.00	0.00	0.00	0.00	0.00	0.00	0.00	0.00	-1.00
AA	S99	0.00	0.00	0.00	0.00	0.00	0.00	0.00	0.00	0.00	0.00	0.00	0.00	0.00	0.00	0.00	-1.00
AA	S99	0.00	0.00	0.00	0.00	0.00	0.00	0.00	0.00	0.00	0.00	0.00	0.00	0.00	0.00	0.00	-1.00
AA	S99	0.00	0.00	0.00	0.00	0.00	0.00	0.00	0.00	0.00	0.00	0.00	0.00	0.00	0.00	0.00	-1.00
AA	S99	0.00	0.00	0.00	0.00	0.00	0.00	0.00	0.00	0.00	0.00	0.00	0.00	0.00	0.00	0.00	-1.00
AA	S99	0.00	0.00	0.00	0.00	0.00	0.00	0.00	0.00	0.00	0.00	0.00	0.00	0.00	0.00	0.00	-1.00
AA	S99	0.00	0.00	0.00	0.00	0.00	0.00	0.00	0.00	0.00	0.00	0.00	0.00	0.00	0.00	0.00	-1.00
AA	S99	0.00	0.00	0.00	0.00	0.00	0.00	0.00	0.00	0.00	0.00	0.00	0.00	0.00	0.00	0.00	-1.00
AA	S99	0.00	0.00	0.00	0.00	0.00	0.00	0.00	0.00	0.00	0.00	0.00	0.00	0.00	0.00	0.00	-1.00
AA	S99	0.00	0.00	0.00	0.00	0.00	0.00	0.00	0.00	0.00	0.00	0.00	0.00	0.00	0.00	0.00	-1.00

Baricenter W: average of the resultant degrees

| | | | | | | | | | | | | | | 0.712 | 0.308 | 0.403 | 0.020 |

It calculate the degree of certainty of the desease AA (H1 = 0.403) before the presented symptoms by the patient

Table 6.21 Decision table, obtained from the symptoms presented by the patient and from the degrees of certainty of each one of the analyzed diseases

1	2	3	4	5	6
Number of presented symptoms = **12**			Certainty degree	Diagnosis prevision	Level of requirement
Possible symptoms	Pres ented symptoms	Possible deseases	H	Most probable desease	**0.60**
S01	S01	AA	0.403		
S02	S02	AB	0.320		
S03	S03	AC	0.165		
S04	S05	AD	0.382		
S05	S07	AE	0.237		
S06	S09	AF	0.651		Acceptable
S07	S12	AG	0.053		
S08	S15	AH	−0.148		
S09	S18	AI	0.386		
S10	S22	AJ	0.564		
S11	S26	AK	0.275		
S12	S30	AL	0.189		
S13	S99	AM	0.398		
S14	S99	NA	0.523		
S15	S99	AO	0.455		
S16	S99	AP	0.728	AP Disease	Acceptable
S17	S99	AQ	0.128		
S18	S99	AR	−0.108		
S19	S99	AS	0.391		
S20	S99	AT	0.701		Acceptable
S21	S99	AU	0.273		
S22	S99	AV	0.170		
S23	S99	AX	0.379		
S24	S99	AY	0.394		
S25	S99	AZ	0.455		
S26	S99	AW	0.709		Acceptable
S27	S99	BA	−0.003		
S28	S99	BB	−0.123		
S29	S99	BC	0.391		
S30	S99	BD	0.696		Acceptable
S99	S99	BE	0.575		
		BF	0.071		

S_{26} and S_{30}. The detection of these symptoms may be performed by the screening sector of the hospital, by means of an interview with the patient.

These symptoms are placed in column 2 of the decision Table 6.19. Column 2, which has 30 lines to receive up to 30 symptoms, must be filled out with the twelve symptoms presented by the patient and completed. For this purpose, besides the symptoms presented by the patient (twelve, in this example), S_{99} must be placed in the other lines.

S_{99} means a total lack of information (paracompleteness) concerning the other possible symptoms, that is, it means that, about all the other symptoms, the values of favorable evidence and contrary evidence are equal to zero.

In fact, if the patient does not present any other symptom, besides the twelve ones verified, these other ones cannot influence the prediction of the diagnosis of its disease.

Besides placing the symptoms presented by the patient, some symptoms (in this case, twelve) must be put in the first line of column 3 of Table 6.20.

Here, in this text, only a fragment of the calculation is shown Table 6.21, that is, only the calculation table referring to disease AA (D_1). However, the CP of the PDM conducts this operation for the thirty-two diseases. The complete calculation table for the thirty-two diseases that are part of the database is in Appendix D, as Sect. 6.5—Text.

Figure 6.15 represents the isolated effects (twelve points) and resultant (center of gravity) of the symptoms presented by the patient in disease AA (D_1). Analogous representations for all the other diseases are in Appendix D, as Sect. 6.5—Text.

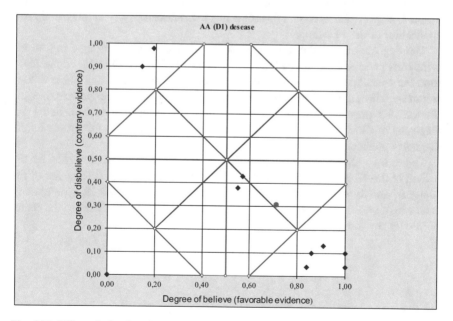

Fig. 6.15 Effects, isolated and resultant, of the twelve symptoms presented by the patient in disease AA (D_1)

6.5.3   The Obtention of the Foreseen Diagnosis

For each disease D_i, the application of the maximizing (**MAX**) and minimizing (**MIN**) operators to each one of the symptoms S_j results in an ordered pair (columns 15 and 16 of Table 6.21). This ordered pair defines in the para-analyzing algorithm a point $X_{i,j} = (a_{i,j}; b_{i,j})$, which translates the effect (for purposes of diagnosing) of symptom j in disease i. The average (arithmetic or weighted) of these annotations defines $W_i = (a_{i,W}; b_{i,W})$ (last line of columns 15 and 16 of Table 6.21), center of gravity of points $X_{i,j}$, which translates the joint effect of the twelve symptoms S_j in disease D_i.

With the annotations of the center of gravity W_i, the CP of the PDM calculates its degree of certainty ($H_{i,W} = a_{i,W} - b_{i,W}$) (last line, column 17, Table 6.21).

For the case of disease AA (disease D_1), whose calculation was shown in 6.21, as a result of the twelve symptoms presented by the patient, the degree of certainty of the center of gravity Wj was calculated the following way:

$$H_{1,W} = a_{1,W} - b_{1,W} = 0.712 - 0.308 = 0.403$$

It must be observed that the prediction process may become more refined, by the attribution of weights to the symptoms, according to their "strength" as an indicator of each one of the related diseases. In this case, the center of gravity would be obtained by the weighted average (and not arithmetic) of the values $a_{i,j}$ and $b_{i,j}$. However, considering that it just concerns performing a prediction and that the idea is to use the result only for the submission of the patient to the specific sector in the hospital, it was considered that this refinement was not necessary in this example of application of the PDM.

The degrees of certainty ($H_{i,W}$) of all the diseases (D_i), resulting from all the symptoms presented by the patient, are "transported" by the own CP of the PDM from the calculation Table 6.21, last line, column 17 to the decision Table 6.20— column 4. Then, the method compares these values with each other and chooses the disease that presents the highest degree of certainty, which is considered the diagnosis prediction. It is shown in column 5 of Table 6.20. In this example, the diagnosis prediction is disease AP.

Figure 6.16 shows the resultant effect of the 12 symptoms presented by the patient in the 32 considered diseases. That is, each point of Fig. 6.16 is one of the center of gravity of Fig. 6.15, which represents the effect of the 12 symptoms in each one of the 32 diseases. They are obtained the same way we obtained the center of gravity referring to disease AA (Fig. 6.15).

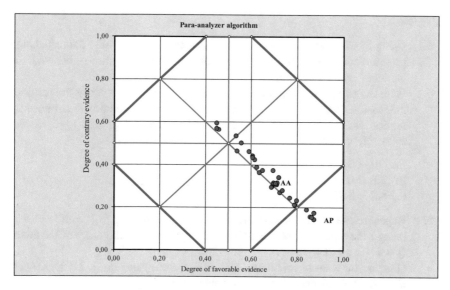

Fig. 6.16 Location of the thirty-two centers of gravity, which translate the combined influence of the twelve symptoms presented by the patient in the thirty-two considered diseases

6.5.4 Restriction to Accept the Foreseen Diagnosis

When performing a disease diagnosis prediction, a requirement may be done about the degree of certainty obtained for each one of them. We may demand that the result is only acceptable if the degree of certainty is higher than a predetermined value. Thus, to be acceptable candidates to the diagnosis, their degrees of certainty must be greater than this minimum. In case none of the diseases meet this requirement, it will be said: there is no diagnosis prediction.

For example, imagine that it is determined that the diagnosis prediction be only accepted is the maximum degree of certainty (from the disease with the highest degree of certainty) is equal or higher than 0.60. Then, this value of 0.60 is adopted as requirement level, once only for values of the degree of certainty equal or higher than it; the prediction will be accepted (the tables and figures of this paragraph were constructed, adopting the requirement level equal to 0.60).

The analysis of the result performed by the para-analyzing algorithm, as shown in Fig. 6.16, provides a clear view of this possible requirement. Thus, a prediction will only be accepted if the center of gravity W_1 of the points that translate the influence of the symptoms on the disease with the highest degree of certainty belongs to the truth region (feasibility).

In the example being analyzed, there are five diseases (AF, AP, AT, AW and BD) (see column 6, Table 6.20) that meet the minimum requirement (NE = 0.60), that is, there are five diseases that are acceptable candidates to be the prediction, but the foreseen diagnosis is disease AP, whose degree of certainty resulted in maximum.

Exercises

6.1   Using the database BD—Sect. 6.1 of Appendix C, assemble the calculation program utilized in this exercise, building the calculation table and the para-analyzing algorithm.

6.2   With the calculation program assembled in exercise 6.1, verify the result of the feasibility analysis when all the factors are in the condition determined by section S_3, that is, when the researched matrix is $[S_j]$ and $j = 3$, in the following cases:

(a) NE $= 0.6$;
(b) NE $= 0.4$;
(c) NE $= 0.8$.

6.3   Repeat exercise 6.2, for the following researched matrix: $[S_j]$, with $j = 1 + r$, being r the remainder of the division of i by 3, being i the order of the factor ($j = 1 + r$ and $i = 3q + r$).
Hint: if you want to calculate r by Excel, use the function: $= MOD(X;3)$ (X must be the cell that contains the value of i)

6.4   Repeat exercises 6.2 and 6.3, attributing different weights to the factors and assuming that the matrix of the weights is $[P_i]$, with $P_i = 5 - 2r$, being r the remainder of the division of i by 3 ($P_i = 5 - 2r$ and $i = 3q + r$).

6.5   Repeat exercise 6.1, assuming that the specialists are distributed in three groups:
A, composed by E_1; B, by E_3 and C, composed by E_2 and E_4.

6.6   With the material assembled in exercise 6.5, repeat exercises 6.2, 6.3 and 6.4.

6.7   Now you play the role of the KE (knowledge engineer): repeat exercise 6.1, choosing only the 8 factors that you deem of most relevance and adopting for them the weights you consider adequate to translate the relative importance of these factors.

6.8   Using the database of Table 6.6, whose electronic version is the database BD—Sect. 6.2 of Appendix C, assemble the CP of the PDM utilized in the considerations of this section.

6.9   Apply the program assembled in exercise 6.8 to analyze the case in which the researched matrix $[S_{pj}] = [S_k]$, being k the biggest integer contained in $(i + 1)/2$ (i is the index of factor F_i), that is, $k = I[(i + 1)/2]$. Determine the center of gravity and its degree of certainty, and verify what is the decision for the requirement levels:

(a) 0.40;
(b) 0.50;
(c) 0.60;

Hint: if you want to calculate the biggest integer $I[(i + 1)/2]$ by Excel, use the function: $= INT(X)$, being $X = (Y + 1)/2$ and Y the cell that contains the value of i.

6.10 Repeat exercise 6.9, for k equal to 1 + r, being r the remainder of the division of i by 3. (k = 1 + r, being i = 3q + r).

6.11 Repeat exercise 6.9, for k equal to 5 − t, being r the remainder of the division of i by 2.

6.12 Solve exercise 6.10, attributing to the factors weights equal to 5 − 2r.

6.13 About exercise 6.12, study the sensitivity of the decision according to the requirement level (NE), elaborating a decision table according to NE, for NE varying from 0.1 to 0.9, with intervals equal to 0.1.

6.14 In order to verify the sensitivity of the decision according to the researched matrix [S_{pj}] and the NE, elaborate a double-entry table for the decision, combining the five researched matrices [S_k], with k = 1, 2, 3, 4 or 5, with the following nine values of NE: from 0.1 to 0.9, varying every 0.1. Adopt the weights used in exercise 6.11. Calculate the degree of certainty of the center of gravity (H_W) for each one of the five values of k.

6.15 Using the database of Table 6.12 (which is under the electronic form in BD —Sect. 6.3 of Appendix C), assemble the CP of the PDM utilized for the obtention of the results presented in Sect. 6.3 (do not forget the matrix of the weights, which may be obtained in Tables 6.13 or 6.14 and 6.15).

6.16 Using the program obtained in exercise 6.15, determine the degrees of evidence, favorable and contrary, and the degree of certainty of the center of gravity, in the case when the transpose of the researched matrix is [S_{pj}]^t = [S_1, S_2, S_3, S_1, S_2, S_3, S_1 S_2].

6.17 Redo exercise 6.16, assuming that the transpose of the matrix of the weights is [P_i]^t = [5, 3, 1, 6, 2, 1, 10, 2].

6.18 Elaborate the decision table suggested by the analysis according to the requirement level (NE) when it assumes the values of the set {0.1; 0.2; 0.3; 0.4; 0.5; 0.6; 0.7; 0.8; 0.9}, in the condition of exercise 6.16 and exercise 6.17.

6.19 To have an idea of the sensitivity of the result of the analysis about the researched sections and to the requirement level, elaborate a double-entry table for the decisions, using the matrix of the weights of exercise 6.17. For this purpose divide the factors into two groups of four, F_1 to F_4 and F_5 to F_8, and adopt that, in each group, the researched sections are equal. This way, the researched matrix may assume nine different compositions. Using these nine different matrices and attributing to the requirement level variable values from 0.1 to 0.9, with intervals of one tenth, build the requested table. Calculate the degree of certainty of the center of gravity in each case.

6.20 Using the database of Table 6.16 (which is under the electronic form in BD —Sect. 6.3 of Appendix C), assemble the CP of the PDM utilized in the calculations of this section.

6.21 Based on the program of exercise 6.20, calculate the degrees of evidence, favorable and contrary, of the center of gravity and its degree of certainty, altering only the researched matrix, so that its transpose changes to [$S_{pj.}$]^t = [S_1 S_1, S_1, S_1, S_2, S_2, S_2, S_2, S_3, S_3, S_4, S_4, S_5, S_5].

6.22 Based on the program of exercise 6.20, calculate the degrees of evidence, favorable and contrary, of the center of gravity and its degree of certainty, altering only the weights of the factors, so that all have equal weights.

6.23 Assume that, for the implementation of the CIM—Computer Integrated Manufacture, the most relevant factors are: F_{01}, F_{04}, F_{05}, F_{08}, F_{15}, F_{16}, F_{21}, F_{22}, F_{26}, F_{27}, F_{29} and F_{31} and that, in the field research, these factors are in the conditions defined by sections S_1, S_1, S_5, S_1, S_2, S_1, S_2, S_4, S_3, S_1, S_1, and S_1, respectively. Considering that the transpose of the matrix of the weights of these factors in the implementation of the CIM is $[Pi]^t = [3, 3, 2, 4, 6, 5, 1, 2, 2, 4, 5, 3]$, perform the feasibility analysis of the implementation of the CIM to the requirement levels
     (a) 0.55 and (b) 0.70.

6.24 Repeat the analysis of exercise 6.23, assuming that all the factors have equal weights.

6.25 Repeat the analysis of exercise 6.23, assuming that the weight of each factor is equal to (I) j; and (II) 6 − j, being j the index of section S_j corresponding to it.

6.26 Assuming that, for the implementation of the GT/CM—Group Technology and Cellular Manufacture, the factors with most influence are F_{12}, F_{14}, F_{16}, F_{17}, F_{19}, F_{22}, F_{23}, F_{25}, F_{29} and F_{31} and that the transpose of the researched matrix is $[S_{pj}]^t = [S_1, S_1, S_1, S_1, S_2, S_2, S_2, S_3, S_3, S_4]$, and also assuming that the transpose of the matrix of the weights is $[P_i]^t = [4, 5, 6, 7, 3, 4, 5, 2, 3, 1]$, analyze the feasibility of this implementation to the significance levels 0.50 and 0.80.

6.27 Repeat exercise 6.26, assuming that the weight of each factor is equal to (I) j; and (II) 6 − j, being j the index of section S_j corresponding to it.

6.28 Calculate the degree of certainty of the center of gravity (H_w), using the weights of exercise 6.26 when all the factors are in section (a) S_1; (b) S_2; (c) S_3; (d) S_4 and (e) S_5.

6.29 Repeat exercise 6.28, considering all the factors with equal weights.

6.30 Suppose that the available data enables us to know in which section only 13 of the 32 factors (indicators) listed in the text. Thus, utilizing the assembly of exercise 6.20 (the same 4 specialists, distributed in groups the same way) and the information in the table below, and based on these 13 indicators, determine which advanced technology system (FMS, CIM, or GT/CM) is the most indicated for these conditions.

Factor Fi		F01	F04	F07	F08	F12	F14	F16	F18	F20	F22	F27	F29	F31
Section Sj		S1	S2	S1	S1	S1	S1	S1	S5	S4	S1	S3	S1	S2
Weight Pi	FMS	5	4	10	5	3	0	0	1	2	0	0	3	0
	CIM	3	3	0	4	0	0	5	0	0	2	4	5	3
	GT/CM	0	0	0	0	4	5	6	0	0	4	0	3	1

6.31 Using the database BD-Ex. 6.31 of Appendix C, which presents the opinions of four specialists E_k ($1 \leq k \leq 4$) related to 10 diseases D_i ($1 \leq i \leq 10$) in the face of 10 symptoms S_j ($1 \leq j \leq 10$), assemble the CP of the PDM for diagnosis prediction. Adopt the requirement level 0.60 to decide if a disease is acceptable or not to be the diagnosis. Adopt group A: E_1 and E_2; group B: E_3 and E_4.

6.32 Utilizing the program assembled in exercise 6.31, in the case when a patient presents the first 5 symptoms S_j ($1 \leq j \leq 5$), verify: (a) which of the 10 diseases are acceptable for diagnosis prediction, at the requirement level 0.60, and their degrees of certainty; (b) what is the foreseen diagnosis.

6.33 Repeat exercise 6.32, for the 5 last symptoms S_j, ($6 \leq j \leq 10$).

6.34 What would the answer to exercise 6.33 be if the requirement level were 0.80?

6.35 If the patient presents the first 3 (1–3) and the last 3 (8–10) symptoms, what will the diagnosis prediction be, at the requirement level of 0.60?

6.36 Solve exercise 6.32, considering that the patient presents the first seven symptoms ($1 \leq j \leq 7$).

6.37 What would the answer to exercise 6.36 be if the requirement level were 0.70?

6.38 Repeat exercise 6.36, considering the last seven symptoms ($4 \leq j \leq 10$).

6.39 Imagine that the patient presents only one symptom at a time. Elaborate a double-entry table, placing the symptoms on the vertical and the diseases on the horizontal. In the intersection of the lines with the columns, place the degree of certainty of the diseases that resulted in diagnosis prediction.

6.40 Repeat exercise 6.39, assuming that the patient presents two symptoms and that these symptoms are the ones obtained from the even symptoms, combining them in pairs.

Answers

6.1 See in Appendix D: Solutions for this chapter, Sect. 6.1—E.g. 6.1

6.2 W = (0.33; 0.73) and H_W = 0.40;
   (a) Inconclusive analysis;
   (b) Non-viable enterprise;
   (c) Inconclusive analysis.

6.3 W = (0.67; 0.52) and H_W = 0.15;
   (a) Inconclusive analysis;
   (b) Inconclusive analysis;
   (c) Inconclusive analysis.

6.4 (6.2) W = (0.33; 0.71) and H_W = −0.38;
   (a) Inconclusive analysis;
   (b) Inconclusive analysis;
   (c) Inconclusive analysis.

6.4 (6.3) W = (0.79; 0.38) and H_W = 0.41;
   (a) Inconclusive analysis;

(b) Viable enterprise;

(c) Inconclusive analysis.

6.6   (6.2) W = (0.27; 0.85) and H_W = −0.58;

(a) Inconclusive analysis;

(b) Inconclusive analysis;

(c) Inconclusive analysis.

6.6   (6.3) W = (0.60; 0.54) and H_W = 0.06;

(a) Inconclusive analysis;

(b) Inconclusive analysis;

(c) Inconclusive analysis.

6.6   (6.4–6.2) W = (0.28; 0.86) and H_W = −0.58;

(a) Inconclusive analysis;

(b) Non-viable enterprise;

(c) Inconclusive analysis.

6.6   (6.4–6.3) W = (0.74; 0.41) and H_W = 0.33;

(a) Inconclusive analysis;

(b) Inconclusive analysis;

(c) Inconclusive analysis.

6.8   See in Appendix D: Solutions for this chapter, Sect. 6.2—E.g. 6.8

6.9   W = (0.55; 0.47) and H_W = −0.08;

(a) Inconclusive;

(b) Inconclusive;

(c) Inconclusive.

6.10   W = (0.74; 0.26) and H_W = −0.48;

(a) Favorable;

(b) Inconclusive;

(c) Inconclusive.

6.11   W = (0.24; 0.79) and H_W = −0.55;

(a) Unfavorable;

(b) Unfavorable;

(c) Inconclusive.

6.12   W = (0.82; 0.19) and H_W = −0.63;

(a) Favorable;

(b) Favorable;

(c) Favorable.

6.13   .

L of R	Decision
0.1	Favorable (viable)
0.2	Favorable (viable)
0.3	Favorable (viable)
0.4	Favorable (viable)

(continued)

(continued)

L of R	Decision
0.5	Favorable (viable)
0.6	Favorable (viable)
0.7	Not conclusive
0.8	Not conclusive
0.9	Not conclusive

6.14 .

L of R	k = 1	k = 2	k = 3	k = 4	k = 5
0.1	Favorable	Favorable	Not Conclusive	Favorable	Favorable
0.2	Favorable	Favorable	Not Conclusive	Favorable	Favorable
0.3	Favorable	Favorable	Not Conclusive	Favorable	Favorable
0.4	Favorable	Favorable	Not Conclusive	Not Conclusive	Favorable
0.5	Favorable	Favorable	Not Conclusive	Not Conclusive	Favorable
0.6	Favorable	Not Conclusive	Not Conclusive	Not Conclusive	Favorable
0.7	Favorable	Not Conclusive	Not Conclusive	Not Conclusive	Favorable
0.8	Favorable	Not Conclusive	Not Conclusive	Not Conclusive	Not Conclusive
0.9	Not Conclusive	Not Conclusive	Not Conclusive	Not Conclusive	Not Conclusive

6.15   See in Appendix D: Solutions for this chapter, Sect. 6.3—E.g. 6.15
6.16   W = (0.62; 0.44) and H_W = 0.18;
6.17   W = (0.76; 0.33) and H_W = 0.43;

6.18 .

L of R	Decision	
	6.16	6.17
0.1	Favorable	Favorable
0.2	Not conclusive	Favorable
0.3	Not conclusive	Favorable
0.4	Not conclusive	Favorable

(continued)

(continued)

L of R	Decision	
	6.16	6.17
0.5	Not conclusive	Not conclusive
0.6	Not conclusive	Not conclusive
0.7	Not conclusive	Not conclusive
0.8	Not conclusive	Not conclusive
0.9	Not conclusive	Not conclusive

6.19 .

L of R	(S1; S1)	(S1; S2)	(S1; S3)	(S2; S1)	(S2; S2)	(S2; S3)	(S3; S1)	(S3; S2)	(S3; S3)
0.1	F	F	NC	F	U	U	NC	U	U
0.2	F	F	NC	F	NC	U	NC	U	U
0.3	F	NC	NC	F	NC	U	NC	U	U
0.4	F	NC	NC	NC	NC	U	NC	U	U
0.5	F	NC	NC	NC	NC	NC	NC	U	U
0.6	F	NC	NC	NC	NC	NC	NC	NC	U
0.7	F	NC	NC	NC	NC	NC	NC	NC	U
0.8	NC	NC	NC	NC	NC	NC	NC	NC	U
0.9	NC	NC	NC	NC	NC	NC	NC	NC	NC
H_w	0.70	0.27	−0.09	0.32	−0.10	−0.47	−0.07	−0.50	−0.86
V = viable of favorable (12 cases)				NC = Not conclusive(51)			U = unviable or contrary (18)		

F favorable (12 situations), NC inconclusive (51) D unfavorable (18)

6.20   See in Appendix D: Solutions for this chapter, Sect. 6.4—E.g. 6.20
6.21   W = (0.72; 0.27) and H_W = 0.45;
6.22   W = (0.73; 0.29) and H_W = 0.44;
6.23   (H_W = 0.59);
     (a) Viable;
     (b) Inconclusive.
6.24   (H_W = 0.48);
     (a) Inconclusive;
     (b) Inconclusive.
6.25   (I) (H_W = 0.17);
     (a) Inconclusive;
     (b) Inconclusive.
     (II) (H_W = 0.63);
     (a) Viable;
     (b) Inconclusive.

6.26 ($H_W = 0.57$);
      (a) Viable;
      (b) Inconclusive.
6.27 (I) ($H_W = 0.14$);
      (a) Inconclusive;
      (b) Inconclusive.
      (II) ($H_W = 0.52$);
      (a) Viable;
      (b) Inconclusive.
6.28 (I)
      (a) 0.82;
      (b) 0.45;
      (c) 0.09;
      (d) −0.62;
      (e) −0.77;
      (II)
      (a) 0.85;
      (b) 0.36;
      (c) 0.09;
      (d) −0.56;
      (e) −0.77;
6.29 (I)
      (a) 0.84;
      (b) 0.39;
      (c) 0.09;
      (d) −0.58;
      (e) −0.77.
6.30 *FMS*: $H_W = 0.74$; *CIM*: $H_W = 0.59$; *GT/CM*: $H_W = 0.87$.
      Therefore, the most indicated system in these conditions is GT/CM.
6.31 See in Appendix D: Solutions for this chapter, Sect. 6.5—E.g. 6.31
6.32 (a) D_1 (with $H_1 = 0.818$) and D_6 (with $H_6 = 0.834$); (b) D_6.
6.33 (a) D_3 (with $H_3 = 0.800$) and D_4 (with $H_4 = 0.814$) and D_{10} (with $H_{10} = 0.794$); (b) D_4.
6.34 (a) D_3 (with $H_3 = 0.800$) and D_4 (with $H_4 = 0.814$); (b) D_4.
6.35 The foreseen diagnosis would be D_{10}, which presents the highest degree of certainty ($H_{10} = 0.543$), but it cannot be accepted, because H_{10} is lower than the requirement level 0.60.
6.36 (a) D_1 ($H_1 = 0.629$), D_5 ($H_5 = 0.617$); D_6 ($H_6 = 0.839$) and D_{10} ($H_{10} = 0.606$); (b) D_6.
6.37 (a) D_6 ($H_6 = 0.839$); (b) D_6.
6.38 (a) D_3 ($H_3 = 0.701$); and D_{10} ($H_{10} = 0.797$); (b) D_{10}.

6.39   .

Sj\Di	D1	D2	D3	D4	D5	D6	D7	D8	D9	D10
S1	0.800					0.800				
S2	0.960					0.960				
S3		0.880				0.880				
S4					0.810				0.810	
S5		0.820								0.820
S6					0.850			0.850		
S7			0.900							0.900
S8				0.790		0.790				
S9				0.800						0.800
S10			0.790						0.790	

6.40   .

Sj\Di	D1	D2	D3	D4	D5	D6	D7	D8	D9	D10
S2 e S4	0.875									
S2 e S6						0.890				
S2 e S8						0.875				
S2 e S10						0.515				
S4 e S6					0.830					
S4 e S8										0.765
S4 e S10									0.800	
S6 e S8				.815						
S6 e S10			0.805							
S6 e S10				0.775						

Chapter 7
Comparison Between the Paraconsistent Decision Method (PDM) and the Statistical Decision Method (EDM)

7.1 An Example to Substantiate the Comparison

The basic idea of this chapter is to make a comparison between the paraconsistent and statistical decision methods.

For the comparison, as an example, we will consider an enterprise in which only ten factors (F_{01}–F_{10}) have considerable influence. We will assume that the opinions of four specialists (E_k) were researched and that, for the application of the maximization (**MAX**) and minimization (**MIN**) rules, they were grouped as follows: Group A: (E_1 with E_2) and Group B: (E_3 with E_4).

This way, the arrangement for application of the **MAX** and **MIN** operators is the following:

$$\textbf{MIN}\{\textbf{MAX}[(E_1), (E_2)], \textbf{MAX}[(E_3), (E_4)]\}$$

For the decision making, the requirement level of 0.70 will be established. That means that we will make a decision if $|H| \geq 0.70$. This way, the decision rule is the following:

> **H ≥ 0.70 ⇒ favorable decision (feasible enterprise);**
>
> **H ≤ − 0.70 ⇒ unfavorable decision (unfeasible enterprise);**
>
> **− 0.70 < H < 0.70 ⇒ inconclusive analysis.**

Table 7.1 shows, in columns 2–9, the degrees of favorable evidence and contrary evidence the specialists attributed to the factors in their real conditions; in columns 10–13, the results of the application of the maximization rule (**MAX**) rule intra-group; in columns 14 and 15, the degrees of favorable evidence ($a_{i,R}$) and contrary evidence ($b_{i,R}$) resultant from the application of the minimization (**MIN**) rule intergroup; and in columns 16–18, the analysis of the results.

© Springer International Publishing AG, part of Springer Nature 2018
F. R. de Carvalho and J. M. Abe, *A Paraconsistent Decision-Making Method*,
Smart Innovation, Systems and Technologies 87,
https://doi.org/10.1007/978-3-319-74110-9_7

Table 7.1 Calculation table

1	2	3	4	5	6	7	8	9	10	11	12	13	14	15	16	17	18
Fi	E1		E2		E3		E4		MAX {E1,E2}		MAX {E3,E4}		MIN {A,B}		L of R=		0.70
	$a_{i,1}$	$b_{i,1}$	$a_{i,2}$	$b_{i,2}$	$a_{i,3}$	$b_{i,3}$	$a_{i,4}$	$b_{i,4}$	$a_{i,gA}$	$b_{i,gA}$	$a_{i,gB}$	$b_{i,gB}$	$a_{i,R}$	$b_{i,R}$	H	G	Decision
F01	0.88	0.04	0.94	0.14	0.84	0.08	0.78	0.03	0.94	0.04	0.84	0.03	0.84	0.04	0.80	-0.12	Viable
F02	1.00	0.05	0.95	0.15	1.00	0.10	0.85	0.00	1.00	0.05	1.00	0.00	1.00	0.05	0.95	0.05	Viable
F03	0.92	0.08	0.98	0.18	0.88	0.12	0.82	0.07	0.98	0.08	0.88	0.07	0.88	0.08	0.80	-0.04	Viable
F04	0.95	0.11	1.00	0.21	0.91	0.15	0.85	0.10	1.00	0.11	0.91	0.10	0.91	0.11	0.80	0.02	Viable
F05	1.00	0.88	0.06	0.10	0.95	0.85	0.04	0.00	1.00	0.10	0.95	0.00	0.95	0.10	0.85	0.05	Viable
F06	0.90	0.10	1.00	0.10	0.90	0.00	1.00	0.00	1.00	0.10	1.00	0.00	1.00	0.10	0.90	0.10	Viable
F07	0.95	0.15	1.00	0.10	0.85	0.00	1.00	0.05	1.00	0.10	1.00	0.00	1.00	0.10	0.90	0.10	Viable
F08	0.98	0.18	0.88	0.12	0.82	0.07	0.92	0.08	0.98	0.12	0.92	0.07	0.92	0.12	0.80	0.04	Viable
F09	1.00	0.21	0.91	0.15	0.85	0.10	0.95	0.11	1.00	0.15	0.95	0.10	0.95	0.15	0.80	0.10	Viable
F10	0.94	0.14	0.84	0.08	0.78	0.03	0.88	0.04	0.94	0.08	0.88	0.03	0.88	0.08	0.80	-0.04	Viable
Baricenter W: averages of the resultant degrees													0.93	0.09	0.84	0.03	Viable

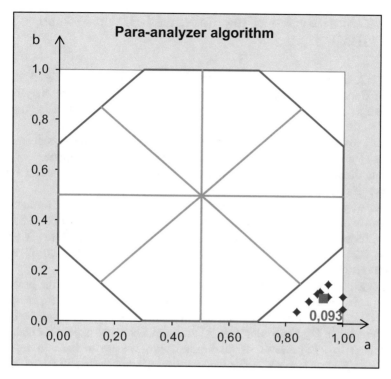

Fig. 7.1 Analysis of the result by the para-analyzing algorithm

The analysis of the results by the decision rule was already executed in Table 7.1. The analysis by the para-analyzing algorithm is performed, as it has already been seen, plotting the resultant degrees of favorable evidence and contrary evidence (columns 14 and 15) in a cartesian diagram and verifying the position of the representative points of each one of the factors and of the center of gravity (Fig. 7.1).

It must be observed that all the factors are in the truth region, which means that all of them are in favorable condition to the enterprise being analyzed.

The influences of all these factors on the decision of the feasibility of the enterprise may be summarized by the center of gravity of the ten points, which translates the joint influence of the ten analyzed factors. As **W** is in the truth region, we say that the analysis enables a favorable decision: the enterprise is feasible.

For the center of gravity, it led to ($H_W = a_W - b_W = 0.84$); as ($0.84 \geq 0.70$), the decision is **favorable**, that is, it is possible to infer for the feasibility of the enterprise.

7.2   A Short Review of the Statistical Decision Method (SDM)

Statistical decisions are the decision making about a population, based on information from the sample(s) extracted from them. For example, we may want to decide if a coin is biased or not, if a drug is more efficient or not than another in the cure of a disease, etc.

To try to reach the decision, statistical hypotheses are formulated concerning the interested population, which are statements regarding the distributions of likelihoods of the population. Commonly, a statistical hypothesis is formulated with the purpose of rejecting it.

This way, when we wish to decide if a coin is biased, we formulate the hypothesis that it is not, that is, that the probability of obtaining one of the faces (heads, for example) is $p = 0.5$. This hypothesis is called **null hypothesis** (H_0: the coin is fair). Any hypothesis other than the null one is denominated **alternative hypothesis** (H_1: $p \neq 0.5$ the coin is not fair, for example).

In practice, we assume H_0 and, based on a random sample and the probability theory; it is verified if the sample results markedly differ from the expected ones, that is if the observed differences are significant to the extent of being able to reject H_0 and maintain H_1. For example, in 50 tosses of a coin, we expect to obtain some heads next to 25; however, if 40 heads occur, we are inclined to reject the hypothesis H_0 that the coin is fair (and keep the alternative hypothesis H_1).

The process which enables us to decide if a hypothesis must be rejected, verifying if the sample data significantly differs from the expected, is called **hypothesis** or **significance test** [99].

If H_0 is rejected when it should be accepted, we say that a **type I** error was made; but if it is accepted when it should be dismissed, the error is **type II**. In both cases, we have a decision error. To reduce these kinds of error, we seek to increase the sample size, which is not always possible.

When testing an established hypothesis, H_0, the maximum probability of making a type **I** error is called **significance level**, generally represented by a and whose most common values are 0.05 (or 5%) and 0.01 (or 1%). Thus, if it is adopted $\alpha = 5\%$ in the planning of the hypothesis test, there are 5 chances in 100 of H_0 being rejected, when it should be accepted, that is, there is a **trust** of 95% of making the right decision. It is said that H_0 is rejected at the **significance level** of 0.05 (or 5%). In the coin example, we would say that there are evidences that the coin is not fair, at the **significance level** of 0.05 (or 5%).

If statistic X has a normal distribution with average p_x and standard deviation μ_x, the distribution of the reduced variable (or score) $z = (X - \mu_x)/\sigma_x$ is normal with average 0 and standard deviation 1.

For the significance level $a = 5\%$, in two-tailed tests, the critical values of z (z_c), which separate the H_0 acceptance region from the rejection region, are -1.96 and $+1.96$. Thus, if the sample result X_o of statistic X leads to a score z_o equal or smaller than -1.96, or equal or higher than $+1.96$, H_0 will be rejected at the significance

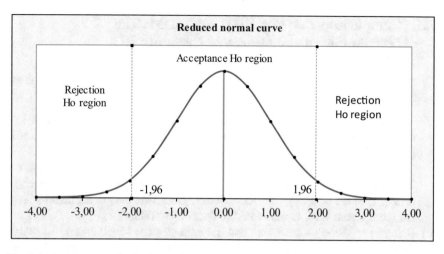

Fig. 7.2 Acceptance and rejection regions in a reduced normal curve, for two-tailed tests with $\alpha = 5\%$

level of 5%. In this case, we say that z_o is significantly different from 0 (average of z) to the extent of enabling us to reject H_0 at the significance level 5%. Therefore, for this significance level, the statistical decision rule is (Fig. 7.2):

Accept H_0 : **if $- 1.96 < z_o < + 1.96$**   or, in a generic way, **if $- z_c < z_o < + z_c$**
Reject H_0 : **if $z_o \leq - 1.96$ or $z_o \geq + 1.96$**   or, in a generic way, **if $z_o \leq - z_c$ or $z_o \geq + z_c$**

For one-tailed tests on the right, for $\alpha = 5\%$, the decision rule is changed to:

Accept H_0 : **if $< + 1.645$**   or, in a generic way, **if $z_o < + z_c$**
Reject H_0 : **if $z_o \geq + 1.645$**   or, in a generic way, **if $z_o \geq + z_c$**

For one-tailed tests on the left, for $\alpha = 5\%$, the decision rule is changed to:

Accept H_0 : **if $< - 1.645$**   or, in a generic way, **if $z_o > - z_c$**
Reject H_0 : **if $z_o \leq - 1.645$**   or, in a generic way, **if $z_o \leq - z_c$**

For the significance level 1%, the critical values of **z** are −2.58 and +2.58 (for two-tailed tests) and −2.33 and +2.33 (for one-tailed tests).

7.3   Comparison Between PDM and SDM: The Distribution of the Degree of Certainty (H)

To compare the paraconsistent Decision Method (PDM) to the statistical decision method (SDM), some considerations about the PDM have been made [51].

(a) The variation interval of the degree of certainty ($-1 \leq H \leq 1$) was divided into classes of amplitude a = 0.1, with extremes in the integer decimal values of H ($0.0 \times 10^{-1}, \pm 10 \times 10^{-1}, \pm 20 \times 10^{-1},\ldots$) (column 2 of Table 7.2).
Thus, the midpoints of the classes are the following: $\pm 0.5 \times 10^{-1} = \pm 0.05, \pm 1.5 \times 10^{-1} = \pm 0.15, \pm 2.5 \times 10^{-1} = \pm 0.25,\ldots, \pm 9.5 \times 10^{-1} = \pm 0.95$. To each class, one integer decimal value of the requirement level is associated (column 1 of Table 7.2).

(b) Being H = M the center of a class, its extremes are M -0.05 and M $+0.05$. Thus, this class is defined by the interval K = M $-0.05 \leq H < M + 0.05$, for H ≥ 0, or M $- 0.05 < H \leq M + 0.05 = K$, for H < 0. K is the requirement level related to the class.

(c) For each class, the area of the Cartesian unit square region defined (delimited) by this class (Fig. 7.3), which was called class area, was calculated. The value $A_M = 0.1 \times (1 - |M|)$ was obtained.

(d) As the Cartesian unit square area is equal to 1, the frequency of the class defined by the value H = M (center of the class) is equal to the class area (A_M) divided by its amplitude (a).
Thus:

$$f_{H=M} = A_M/a = 0.1 \times (1 - |M|)/0.1 = 1 - |M|$$

(e) This way, it was possible to calculate the areas A_M and the frequencies f_H of all the classes (columns 4 and 5 of Table 7.2) and elaborate the corresponding frequencies diagram (Fig. 7.4).

(f) When we assume the requirement level NE = K for the decision making by the PDM, it means that the decision will be favorable if $H_W \geq K$, and unfavorable if $H_W \leq -K$, being H_W the degree of certainty of the center of gravity.

Therefore, the decision is favorable is the center of gravity W belongs to the QUPC region defined by the condition H $\geq$ K, that is, if it belongs to the tail of the curve constituted by the classes of mid-points M, so that M $\geq$ K + 0.05 or $|M| \geq K + 0.05$.

Analogously, the decision is unfavorable is the center of gravity W belongs to the QUPC region defined by the condition H $\leq -K$, that is, if it belongs to the tail of the curve constituted by the classes of mid-points M, so that M $\leq -K - 0.05$ or $|M| \geq K + 0.05$.

Therefore, if the center of gravity W belongs to one of the tails (right or left) of the frequency distribution of H defined by the requirement level NE = K, it means

Table 7.2 Classes, observed (PDM) and expected (normal) frequencies, χ^2 (chi-square) calculation and accumulated areas under the PDM and normal curves

1	2	3	4	5	6	7	8	9	9
Level of requirement (K)	Classes	Midpoint (M)	AM	$f_H = f_o$	$f_N = f_E$	$\dfrac{(f_o - f_E)^2}{f_E}$	A_{acum} MPD	A_{acum} Normal	Corrected A_{acum} Normal
0.9	−1.0 ⊢ −0.9	−0.95	0.005	0.05	0.091	0.01853	0.005	0.009	0.021
0.8	−0.9 ⊢ −0.8	−0.85	0.015	0.15	0.144	0.00027	0.020	0.023	0.035
0.7	−0.8 ⊢ −0.7	−0.75	0.025	0.25	0.216	0.00544	0.045	0.045	0.057
0.6	−0.7 ⊢ −0.6	−0.65	0.035	0.35	0.308	0.00581	0.080	0.076	0.088
0.5	−0.6 ⊢ −0.5	−0.55	0.045	0.45	0.417	0.00258	0.125	0.118	0.130
0.4	−0.5 ⊢ −0.4	−0.45	0.055	0.55	0.538	0.00029	0.180	0.171	0.183
0.3	−0.4 ⊢ −0.3	−0.35	0.065	0.65	0.659	0.00011	0.245	0.237	0.249
0.2	−0.3 ⊢ −0.2	−0.25	0.075	0.75	0.767	0.00037	0.320	0.314	0.326
0.1	−0.2 ⊢ −0.1	−0.15	0.085	0.85	0.849	0.00000	0.405	0.399	0.411
0.0	−0.1 ⊢ 0.0	−0.05	0.095	0.95	0.893	0.00366	0.500	0.488	0.500
0.0	0.0 ⊢ 0.1	0.05	0.095	0.95	0.893	0.00366	0.595	0.577	0.589
0.1	0.1 ⊢ 0.2	0.15	0.085	0.85	0.849	0.00000	0.680	0.662	0.674
0.2	0.2 ⊢ 0.3	0.25	0.075	0.75	0.767	0.00037	0.755	0.739	0.751
0.3	0.3 ⊢ 0.4	0.35	0.065	0.65	0.659	0.00011	0.820	0.805	0.817
0.4	0.4 ⊢ 0.5	0.45	0.055	0.55	0.538	0.00029	0.875	0.858	0.870
0.5	0.5 ⊢ 0.6	0.55	0.045	0.45	0.417	0.00258	0.920	0.900	0.912
0.6	0.6 ⊢ 0.7	0.65	0.035	0.35	0.308	0.00581	0.955	0.931	0.943
0.7	0.7 ⊢ 0.8	0.75	0.025	0.25	0.216	0.00544	0.980	0.953	0.965
0.8	0.8 ⊢ 0.9	0.85	0.015	0.15	0.144	0.00027	0.995	0.967	0.979
0.9	0.9 ⊢ 1.0	0.95	0.005	0.05	0.091	0.01853	1.000	0.976	0.988

$$\chi^2 = 0.07412$$

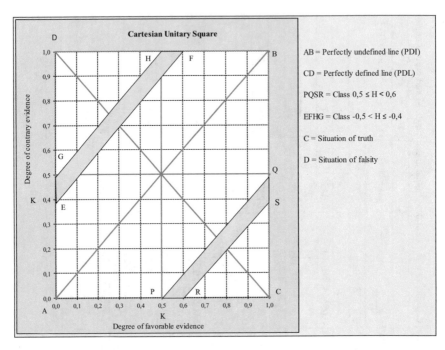

Fig. 7.3 Classes of the degree of certainty, with two corresponding classes to the requirement levels 0.5 and 0.4 highlighted

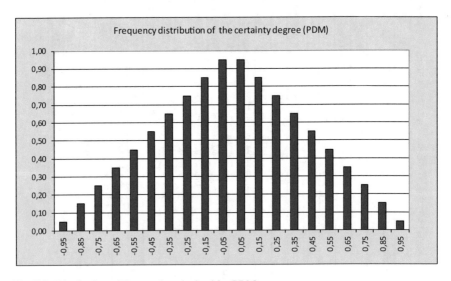

Fig. 7.4 Distribution of frequencies obtained by PDM

that the degree of certainty of the center of gravity is significantly different from zero to be able to make the decision (favorable or unfavorable).

7.4 Comparison Between PDM and SDM: The Adherent Normal Curve (ANC)

To make the comparison between the PDM and the statistical decision process, we sought the normal distribution of zero average (as the distribution of H has zero average) which better adhered to the H frequency distribution (of the PDM).

To measure this adherence, the adherence test of χ^2 (chi-square) was applied. For this purpose, the corresponding frequency to each class of the degree of certainty ($f_0 = f_H$) (column 5 of Table 7.2 and Fig. 7.4) was considered as observed frequency, and the frequency of the same class obtained by the normal curve ($f_E = f_N$) (column 6 of Table 7.2 and Fig. 7.5) as expected frequency. This frequency was obtained with the aid of the Excel spreadsheet, utilizing the function DIST.NORM(X; MEDIA; DESVPAD; FALSO).

It was verified that the best adherence of the normal distribution of zero average to the distribution of the degree of certainty of the PDM occurs for the standard deviation equal to 0.444, for which resulted in a chi-square minimum and equal to $\chi^2 = 0.07412$ (column 7 of Table 7.2 and Fig. 7.6a, b). This normal will be called normal adherent curve (ANC).

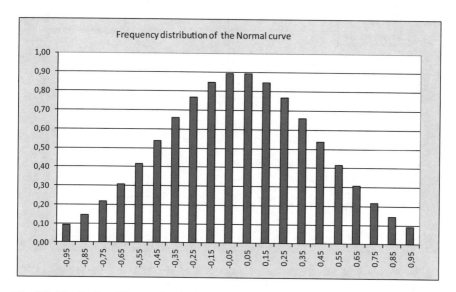

Fig. 7.5 Distribution of frequencies obtained by normal curve of zero average and 0.444 standard deviations

(a)

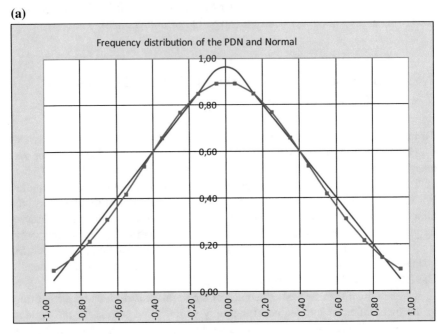

(b)

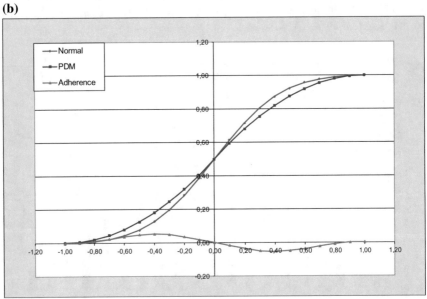

Fig. 7.6 a Curves of the distribution of frequencies of H (PDM) and the normal (ANC). **b** Accumulated frequencies of the distribution of the PDM and of the normal (ANC)

Table 7.3 Comparison between the areas of the tails of the distribution of H (PDM) and the normal (ANC) and the variation in the value of χ^2 for some standard deviation values

Level of requirement	Level of uncertainty	Level of significancy	Standard deviation	χ^2 (chi-square)
Acceptable value minimum of certainty degree	Tail of PDM curve	Tail of normal curve	0.437	0.07683
			0.438	0.07607
			0.439	0.07545
K	β (%)	λ (%)	0.440	0.07494
0	50.00	50.00	0.441	0.07456
0.1	40.50	41.07	0.442	0.07429
0.2	32.00	32.59	0.443	0.07415
0.3	24.50	24.92	*0.444*	*0.07412*
0.4	18.00	18.33	0.445	0.07420
0.5	12.50	12.96	0.446	0.07440
0.6	8.00	8.78	0.447	0.07470
0.7	4.50	5.71	0.448	0.07511
0.8	2.00	3.55	0.449	0.07563
0.9	0.50	2.11	0.450	0.07625

In these conditions, the decision by the PDM with requirement level equal to K (favorable, if $H_W \geq K$, or unfavorable if $H_W \leq -K$) corresponds to a two-tailed statistical decision with significance level $\alpha = 2\lambda$, being λ equal to the area under ANC, above K (favorable decision) or below—K (unfavorable decision) (see Table 7.3 and Fig. 7.7).

It is worth to observe that, for the normal, the area of each class was calculated by the product of its frequency (column 6 of Table 7.2) by the amplitude of the classes (a = 0.1). The accumulated areas of the distributions of H (PDM) and normal (columns 8 and 9 of Table 7.2) were obtained by the accumulated sum of the class areas. In this calculation, for the normal, the corresponding correction to the area under the curve until the value -1.0 was made, obtaining the corrected accumulated area: $A_{\text{acum corrig}}$ (column 10).

For two-tailed tests, the significance level is $\alpha = 2\lambda$; for one-tailed tests, it is $\alpha = \lambda$.

7.5 Comparison Between PDM and SDM: Comparing the Decisions

The double of the area of the normal curve tail (2λ) is called significance level (for two-tailed testing) and represents the percentage of uncertainty with which we accept the decision that the obtained result (H_W) sufficiently different from zero (average of H) to say that the enterprise is feasible (favorable decision) or infeasible (unfavorable decision).

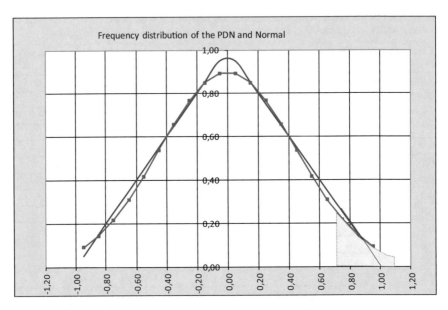

Frequency distribution of the PDN and Normal

Fig. 7.7 Tail of the adherent normal curve (ANC) to the curve of H (PDM

Analogously, the double of the PDM curve tail (2β), which will be called uncertainty level, represents the area of the QUPC region for which $H \geq K$ or $H \leq -K$. Thus, when we say that a decision was made by the PDM with requirement level K, it corresponds to saying that the degree of certainty of the center of gravity, in module, is higher or equal to the requirement level $(|H| > K)$ or that the decision presents a degree of uncertainty 2β.

As seen, to decide the PDM we calculate the degree of certainty of the center of gravity (H_W) and compare it to the requirement level. In the example, we obtained $H_W = 0.84$, which is compared to the requirement level NE = 0.70. As $H_W \geq NE$, we conclude that the decision is favorable (the enterprise is feasible) at the requirement level 0.70. That is, we may state that the company is feasible with maximum uncertainty level of $2\beta = 2 \times 4.50\% = 9.0\%$ (see Table 7.3).

To perform the decision making by the statistical process:

(a) The critical value of the standardized variable of the adherent normal curve ANC $(^*z_c)$ is calculated, which corresponds to the adopted requirement level (0.70, in the example). For this purpose, we verify how many standard deviations of ANC (0.444) the requirement level is above the average (zero), as follows:

$$^*z_c = (0.70 - 0)/0.444 = 1.58$$

(b) We calculate the observed value of the standardized variable of the ANC $(^*z)$ that corresponds to the value of the degree of certainty of the center of gravity (0.840, in the example). For this purpose, we verify how many standard deviations of ANC (0.444) the degree of certainty of the center of gravity is above the average (zero), as follows:

$$^*z_o = (0.840 - 0)/0.444 = 1.89$$

As $^*z_o \geq {}^*z_c$, we conclude that the value H_W is significantly higher than the average zero, enabling us to say that the conclusion of the conducted analysis is favorable (the enterprise is feasible) at the significance level $\alpha = 2\lambda = 2 \times 5.71\% = 11.42\%$ (see Table 7.3).

Note: According to Table 7.3. we notice that, for these one-tailed tests, if the requirement level adopted by the PDM is 0.60, the uncertainty level of the PDM will be 8.00% and the significance level of the SDM will be 8.78%; analogously, if the requirement level is 0.80, these values will be 2.00 and 3.55%, respectively. For two-tailed tests, the values of the uncertainty level and of the significance level double their value.

7.6 Another View of the Application of the Statistic

As already seen, the comparison with the statistic was made only in the final decision, when the value of the degree of certainty of the center of gravity (H_W) was already known. However, assuming that the variable "degree of evidence" may be treated at the level of a reasonable scale, there is also the possibility to use the statistic to work the data and obtain the degree of certainty of the center of gravity.

For each factor F_i, we calculate the average of the degrees of favorable evidence ($a_{i,k}$) related to the opinions of all the experts, obtaining the resultant degree of favorable evidence for this factor ($a_{i,R}$). Analogously, we obtain ($b_{i,R}$). From these values, the process is repeated.

This way, the calculation Table 7.1 is modified and turned into Table 7.4. It must be observed that columns 10–13 of Table 7.1 cease to exist.

As the requirement level is the same, the value of $^*z_c$ remains equal to 1.58.

The degree of certainty of the center of gravity, calculated by the statistic techniques, is now $H_W = 0.743$, which enables us to calculate $^*z_o = (0.743 - 0)/0.444 = 1.67$. As it resulted $^*z_o \geq {}^*z_c$, we conclude that the decision is favorable, that is, the enterprise is feasible, at the significance level $\alpha = 2\lambda = 2 \times 5.71\% = 11.42\%$, which corresponds to the requirement level 0.70.

Then, the doubt arises: the decision by the PDM is with uncertainty level $2\beta = 9.0\%$, corresponding to requirement level 0.70. If we establish the significance level $\alpha = 2\lambda = 9.0\%$, would the decision by the statistic be the same? Just verify which value of the requirement level must be so that the tail of the ANC is equal to 4.50%.

After the verification is conducted, it was obtained that the requirement level that meets is $K' = 0.727$, for which $^*z_c = 1.64$. Therefore, as $^*z_o = 1.67$, the decision is still favorable at the significance level $\alpha = 2\lambda = 2 \times 4.50\% = 9.0\%$, as $^*z_o$ is still higher than $^*z_c$.

Table 7.4 Calculation table of the resultant degrees by means of assigned degrees by the specialists

1	2	3	4	5	6	7	8	9	14	15	16	17	18
F_i	E_1		E_2		E_3		E_4		Degree averages		Level of requirement		0.70
	$a_{i,1}$	$b_{i,1}$	$a_{i,2}$	$b_{i,2}$	$a_{i,3}$	$b_{i,3}$	$a_{i,4}$	$b_{i,4}$	$a_{i,R}$	$b_{i,R}$	H	G	Decision
F_{01}	0.88	0.04	0.94	0.14	0.84	0.08	0.78	0.03	0.86	0.07	0.79	−0.07	Viable
F_{02}	1.00	0.05	0.95	0.15	1.00	0.10	0.85	0.00	0.95	0.08	0.88	0.03	Viable
F_{03}	0.92	0.08	0.98	0.18	0.88	0.12	0.82	0.07	0.90	0.11	0.79	0.01	Viable
F_{04}	0.95	0.11	1.00	0.21	0.91	0.15	0.85	0.10	0.93	0.14	0.79	0.07	Viable
F_{05}	1.00	0.88	0.06	0.10	0.95	0.85	0.04	0.00	0.51	0.46	0.05	−0.03	Viable
F_{06}	0.90	0.10	1.00	0.10	0.90	0.00	1.00	0.00	0.95	0.05	0.90	0.00	Viable
F_{07}	0.95	0.15	1.00	0.10	0.85	0.00	1.00	0.05	0.95	0.08	0.88	0.02	Viable
F_{08}	0.98	0.18	0.88	0.12	0.82	0.07	0.92	0.90	0.90	0.11	0.79	0.01	Viable
F_{09}	1.00	0.21	0.91	0.15	0.85	0.10	0.95	0.93	0.93	0.14	0.79	0.07	Viable
F_{10}	0.94	0.14	0.84	0.08	0.78	0.03	0.88	0.04	0.86	0.07	0.79	−0.07	Viable
Baricenter W: averages of the resultant degrees									0.874	0.131	0.743	0.005	Viable

Of course, the decision could cease being favorable, once the requirement was increased when the significance level was reduced from $\alpha = 2\lambda = 11.42\%$ to $\alpha = 2\lambda = 9.0\%$.

Exercises

7.1   Utilizing the CP of the PDM constructed in Exercise 6.1, and considering the weights of the factors equal, perform the study of the decision by the statistical method, utilizing a two-tailed test, at the significance level $\alpha = 2\lambda = 17.56\%$ (which corresponds to the requirement level 0.60 of the PDM—see Table 7.3), in the cases when all the factors are in the conditions of the section:

$$\text{(a) } S_1 \quad \text{(b) } S_2 \quad \text{(c)} S_3$$

Use the normal curve (ANC) that best adheres to the degree of certainty curve: average 0 (zero) and standard deviation 0.444, and determine the resultant degrees by the average of the degrees attributed by the experts (item 7.6).

Data: for two-tailed test, at the significance level 17.56%, $z_c = 1.354$.

7.2   Compare the decisions made in Exercise 7.1 to the corresponding decisions made by the paraconsistent decision method, PDM, at the requirement level 0.60.

7.3   Which would the answers of Exercise 7.1 be, if the significance level I) $\alpha = 2\lambda = 11.42\%$ (which corresponds to NE = 0.70) and II) $\alpha = 2\lambda = 7.10\%$ (which corresponds to NE = 0.80) were adopted?

Data: for two-tailed test: $\alpha = 2\lambda = 11.42\% \Rightarrow z_c = 1.580$; $\alpha = 2\lambda = 7.10\% \Rightarrow z_c = 1.805$.

7.4   Utilizing the CP of the PDM constructed in Exercise 6.8, and considering the weights of the factors equal, perform the study of the decision by the statistical method, utilizing a two-tailed test, at the significance level $\alpha = 2\lambda = 5.0\%$ (which corresponds to the requirement level 0.843 of the PDM—see Table 7.3), in the cases when all the factors are in the conditions of the section:

$$\text{(a) } S_1 \quad \text{(b) } S_2 \quad \text{(c)} S_5$$

Data: for two-tailed test, at the significance level $\alpha = 2\lambda = 5.0\%$, $z_c = 1.96$.

7.5   Compare the decisions made in Exercise 7.4 to the corresponding decisions made by the paraconsistent decision method, PDM, at the requirement level 0.843.

7.6   Utilizing the CP of the PDM constructed in Exercise 6.15 and adopting the conditions of Exercise 6.19, elaborate the double-entry table for the decisions, highlighting in the first column the requirement level, in the second one, the corresponding significance level for two-tailed test (see Table 7.3), in the third, the critical value of the reduced variable (z_c, which must be obtained in a value table of the normal curve area) and in the nine following ones, the decision suggested by the statistical analysis.

Also, place three last lines with the degrees of certainty of the center of gravity obtained by the PDM (already obtained in Exercise 6.19) and by the statistical calculation, and the observed values of the reduced variable (z_o), corresponding to the degrees of certainty achieved by the statistical calculation. (Do not forget the weights, which are the ones from Exercise 6.17).

Answers

7.1 (a) $z_o = 1.619$; rejects H_0, as $z_o \geq z_c$; Feasible enterprise;

(b) $z_o = 0.605$; accepts H_0, as $-z_c < z_o < z_c$; Inconclusive analysis;

(c) $z_o = -1.154$; accepts H_0, as $-z < z < z$; Inconclusive analysis.

7.2. (a) $H_W = 0.67$; $H_W \geq NE$; Feasible enterprise;

(b) $H_W = 0.22$; $-NE < H_W < NE$; Inconclusive analysis;

(c) $H_W = 0.40$; $-NE < H_W < NE$; Inconclusive analysis;

7.3 (I) (a) $z_o = 1.619$; rejects H_0, as $z_o \geq z_c$; Feasible enterprise;

(b) $z_o = 0.605$; accepts H_0, as $-z_c < z_o < z_c$; Inconclusive analysis;

(c) $z_o = -1.154$; accepts H_0, as $-z_c < z_o < z_c$; Inconclusive analysis;

(II) (a) $z_o = 1.619$; accepts H_0, as $-z_c < z_o < z_c$; Inconclusive analysis;

(b) $z_o = 0.605$; accepts H_0, as $-z_c < z_o < z_c$; Inconclusive analysis;

(c) $z_o = 1.154$; accepts H_0, as $-z_c < z_o < z_c$; Inconclusive analysis;

7.4 (a) $z_o = 1.672$; accepts H_0, as $-z_c < z_o < z_c$; Inconclusive analysis;

(b) $z_o = 0.109$; accepts H_0, as $-z_c < z_o < z_c$; Inconclusive analysis;

(c) $z_o = 1.506$; accepts H_0, as $-z_c < z_o < z_c$; Inconclusive analysis.

7.5 (a) $H_W = 0.840$; $-NE < H_W < NE$; Inconclusive analysis;

(b) $H_W = 0.509$; $-NE < H_W < NE$; Inconclusive analysis;

(c) $H_W = -0.757$; $-NE < H_W < NE$; Inconclusive analysis.

7.6

L of R	L of S (α)	z_c	(S1; S1)	(S1; S2)	(S1; S3)	(S2; S1)	(S2; S2)	(S2; S3)	(S3; S1)	(S3; S2)	(S3; S3)
0.10	82.14%	0.225	F	F	F	F	NC	D	NC	D	D
0.20	65.18%	0.451	F	F	NC	F	NC	D	NC	D	D
0.30	49.84%	0.677	F	F	NC	F	NC	NC	NC	D	D
0.40	36.66%	0.903	F	F	NC	NC	NC	NC	NC	NC	D
0.50	25.92%	1.128	F	NC	NC	NC	NC	NC	NC	NC	D
0.60	17.56%	1.354	F	NC	NC	NC	NC	NC	NC	NC	D
0.70	11.42%	1.580	F	NC	NC	NC	NC	NC	NC	NC	NC
0.80	7.10%	1.805	NC	NC	NC	NC	NC	NC	NC	NC	NC
0.90	4.22%	2.032	NC	NC	NC	NC	NC	NC	NC	NC	NC
By PDM	H_W	0.70	0.27	−0.09	0.32	−0.10	−0,47	−0.07	−0,50	−0.86
By EDM	H_W	0.74	0.41	0.11	0.38	0.05	−0,26	−0,03	−0,36	−0.66
Observed value	z_o	1.667	0.923	0.248	0.856	0.113	−0,586	−0,068	−0,811	−1.486
F = viable or favorable (15 cases)			NC = Not Conclusive (55)				D = unviable or contrary (11)				

Chapter 8
A Simplified Version of the Fuzzy Decision Method and Its Comparison to the Paraconsistent Decision Method

8.1 Simplified Version of the Fuzzy Decision Method (SVFDM)

8.1.1 Theoretical Basis

The inventor of the Fuzzy Logic, in 1965, was the Iranian living in the United States, Lotfi Asker Zadeh. In a less strict language, we may say that this logic seeks a systematization of the study of knowledge, seeking mainly, to examine the vague (diffuse, nebulous) knowledge (it is not clearly known what it means), and distinguish it from inaccurate knowledge (you know what it means, but you do not know the exact value).

Consider X a set (in the usual sense). We say that A is a Fuzzy subset of X, if A is identified by a function $f(x)$ that, at every element of X, associates a number from the interval [0, 1],

If $Y = f(x)$ and [0, 1], for $\forall x \in X$, we denote: $x \in_{f(x)} A$ or $x \in_Y A$. We have:
x belongs to A with membership degree $f(x) = Y$;
$Y = f(x)$ is the membership degree of x in A;
$x \in_Y A$ means that x **belongs** to A with a membership degree Y;
$x \in_0 A$ means that x **does not belong** absolutely to A (it is the case in which the membership degree is $Y = 0$);
$x \in_1 A$ means that x **belongs** absolutely to A (it is the case in which the membership degree is $Y = 1$);
$x \in_{0.7} A$ means that x **belongs** to A with membership degree 0.7 (in this case $Y = 0.7$).

An example with a little more detail. Consider $X = \{a, b, c\}$. Consider the subsets of X with the following elements:

Subset A: $a \in_{0.8} A$, $b \in_{0.3} A$ and $c \in_1 A$. We say that A is a fuzzy subset of X;

© Springer International Publishing AG, part of Springer Nature 2018
F. R. de Carvalho and J. M. Abe, *A Paraconsistent Decision-Making Method*,
Smart Innovation, Systems and Technologies 87,
https://doi.org/10.1007/978-3-319-74110-9_8

Subset B: $a \in_1 B$, $b \in_0 B$ and $c \in_1 B$. It is said that B is a classical subset of X ($B = \{a, c\}$). Therefore, every classical subset of X is a fuzzy subset of X (the elements of the classical subset have membership degree equal to 1).

For the fuzzy subsets of a set X, we define:

Equality: $A = B \Leftrightarrow f_A(x) = f_B(x)$, $\forall x \in X$
Inclusion: $A \subseteq B \Leftrightarrow f_A(x) < f_B(x)$, $\forall x \in X$
Intersection: $C = A \cap B \Leftrightarrow f_c(x) = \min \{(f_A(x), f_B(x)\}$, $\forall x \in X$
Union: $C = A \cup B \Leftrightarrow f_c(x) = \max \{f_A(x), f_B(x)\}$, $\forall x \in X$
Being $x \in_{f(x)} A$ a formula and representing it by $[A(x)] = [A] = f(x) = Y$.

We define

Conjunction: $[A \wedge B] = \max \{[A], [B]\}$
Disjunction: $[A \vee B] = \min \{[A], [B]\}$
Negation: $[A] = 1 - [A]$
Implication: $[A \rightarrow B] = [A \vee B] = \min \{(1 - [A]), [B]\}$.

As an example, analyze the proposition "John is a tall man". In this case, set X is the set of the men. Subset A under analysis is the subset of the tall men (fuzz subset of X), defined by the function f(x) which, at every element x (man) of X, associates a number from the interval [0, 1].

In Classical Logic, he is either tall (membership degree 1) or not tall (membership degree 0) (principle of the middle excluded); that is, x belongs to A or x does not belong to A.

Thus, this logic would define as being "tall", for example, the man with h 1.75 m. The value of function f(x) would be equal to zero for a man x with a height lower than 1.75 m, and it would be equal to 1 for a man x with a height equal or higher than 1.75 m. The interval [0, 1] would be reduced to the binary set {0, 1}.

As the concept of a tall man is a little vague (nebulous, diffuse, fuzzy), the Fuzzy Logic gives it a different treatment. We may consider, for example, that the man is tall for heights above or equal to 1.90 m and not tall for heights below 1.80 m.

Then: for $h \geq 1.90$ m, $f(x) = 1$ and man x absolutely belongs to the fuzzy subset A (of the tall men) of set X (of the men) ($x \in_1 A$); for $h < 1.80$ m, $f(x) = 0$ and man x absolutely does not belong to the fuzzy subset A (of the tall men) of set X (of the men) ($x \in_0 A$) (Fig. 8.1).

However, for 1.80 m $\leq h < 1.90$ m, the value of f(x) will vary from 0 to 1, for example, as follows: $f(x) = 10h_x - 18$ (it was placed, for example, because the function does not need, necessarily, to be linear). Therefore, for a man x of height $h_x = 1.84$ m, we will have $f(x) = 0.4$ and we may say that he belongs to the fuzzy subset A (of the tall men) of set X (of the men), with membership degree 0.4 ($x \in_{04} A$). Figure 8.2 represents the variation of the membership degree.

In short, in this case, the membership function is: $f(x) = 0$, for $h < 1.80$ m; $f(x) = 1$, for $h \geq 1.90$ m; and $f(x) = 10h - 18$, for 1.80 m $\leq h < 1.90$ m.

On the other hand, we may analyze what a "short" man is. As seen above, according to the classical logic, the short man (or not tall) would be the one with

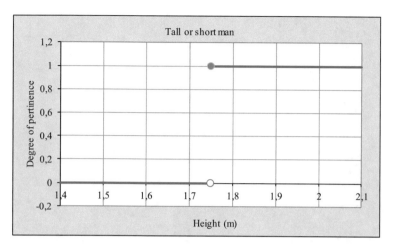

Fig. 8.1 Classical representation of the set of the men (X) and its subset (A) of the tall men

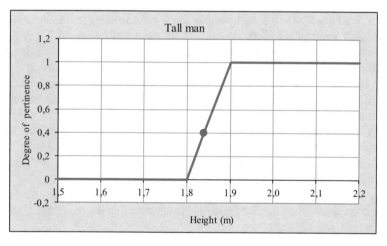

Fig. 8.2 Fuzzy representation of the set of the men (X) and its subset (A) of the tall men

height below 1.75 m. However, under the light of the fuzzy logic, we may say that a short man is the one with a height equal or below 1.60 m and not short the one with height above 1.70 m.

Analogously, we have, then: for $h \leq 1.60$ m, $f(x) = 1$, and man x absolutely belongs to the fuzzy subset B (of the short men) of set X (of the men) ($x \in_0 B$); for $h > 1.70$ m, $f(x) = 0$ and man x absolutely does not belong to the fuzzy subset B (of the short men) of set X (of the men) ($x \in_0 B$).

However, for 1.60 m $< h \leq 1.70$ m, the value of $f(x)$ will vary from 1 to 0, for example, as follows: $f(x) = 17 - 10h_x$. Therefore, for a man x of height $h_x = $

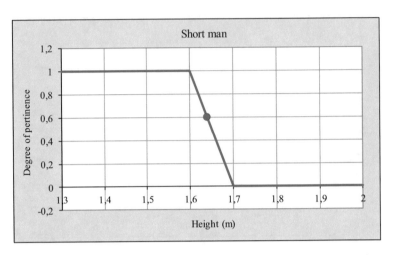

Fig. 8.3 Fuzzy representation of the set of the men (X) and its subset (B) of the tall men

1.64 m, we have f(x) = 0.6, and we may say that he belongs to the fuzzy subset B (of the short men) of set X (of the men), with membership degree 0.6 (x $\in_{0.6}$ B).

Here, the membership function is: f(x) = 0, for h > 1.70 m; f(x) = 1, for h < 1.60 m; and f(x) = 17 − 10 h, for 1.60 m < h ≤ 1.70 m. (Fig. 8.3)

For this work, adaptation will be made, considering that, for a condition (the favorable one), the membership degree varies from 0 to 1, and for the contrary condition (the unfavorable one), it ranges from 0 to −1.

This way, in the example seen above, the man x may belong to subset A of the tall men (here considered as favorable condition) with membership degree varying from 0 to 1; and may belong to subset B of the short men (here considered as unfavorable condition) with membership degree (adapted) from 0 to −1. Observe that, in the interval, 1.70 m ≤ h ≤ 1.80 m, the man will be in subset A of the tall men and the subset B of the short ones with a membership degree (zero). That is, he is not considered tall nor short.

With this adaptation, the fuzzy charts seen are summarized into the one in Fig. 8.4, and the membership function adapted to subset A of the tall men and to subset B of the short men changes to: f(x) = −1, for h ≤ 1.60 m; f(x) = 10 h − 17, for 1.60 m < h ≤ 1.70 m; f(x) = 0, for 1.70 m < h < 1.80 m; f(x) = 10 h − 18, for 1.80 m ≤ h < 1.90 m; and f(x) = 1, for h ≥ 1.90 m.

This way, the advantage of being tall grows until 1.90 m, and from that point on, does not increase any more, remaining equal to 1. That will lead to an important condition in the decision making. Imagine if you want to implement the project of a factory and that this plant needs a minimum area of 4000 m^2 and maximum area of 6000 m^2. Thus, the membership function varies from 0 to 1 for these extremes. However, if the area is larger than 6000 m^2, that will not pose an additional advantage, because the maximum necessary has already been accomplished. Coherently, the membership degree will not increase either, remaining equal to 1.

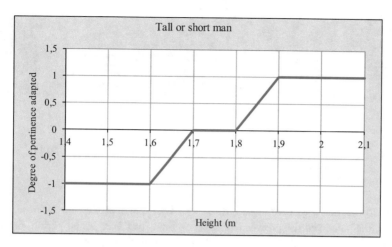

Fig. 8.4 Adapted membership degrees for the subsets of the tall men (A) and the short men (B)

8.1.2   Application of the Fuzzy Logic in Decision Making

One way to utilize the Fuzzy Logic for decision making is the one exposed below, and it will be called simplified version of the fuzzy decision process (SVFDM).

The factors that influence the success (or failure) of an enterprise are considered and, for each factor separately, two ranges are established (by means of specialists), which must translate the condition in which the factor is favorable to the success of the enterprise (RF) and the condition in which it is contrary (RC). Thus, the real state of each factor F_i is translated by the membership degree (Y) of the actual situation of the factor at the established condition. The value of this membership degree will be from 0 to 1, if the actual condition is favorable, and from 0 to −1 if it is contrary. When the membership degree results equal to 0 (zero), it is said that the factor is in an **indifferent** condition for the result of the enterprise.

This way, for each influence factor, we have a membership degree (Y), which translates the real situation. The average (Y_R), (arithmetic or weighted) of all the membership degrees (Y) translates the joint influence of all the factors on the enterprise.

With the value of this average (Y_R), upon a pre-established criterion (decision rule), we may conclude if the enterprise is feasible or infeasible, or if the analysis was inconclusive.

The decision rule is fixed, establishing a minimum value for $|Y_R|$, which will be called **requirement level** (NE). It may also be set by specialists in the area, or its value must depend on the responsibility the decision implicates. Thus, the decision rule may be written as follows:

YR $\geq$ NE $\Rightarrow$ favorable decision (feasible enterprise);
YR $\leq$ −NE $\Rightarrow$ unfavorable decision (unfeasible enterprise);
−NE < YR < NE $\Rightarrow$ inconclusive analysis.

8.1.3   A Simple Application of the SVFDM (Negoita [85] with Shimizu [94])

Eight companies are considered (from A to H), of which we know, for a certain period, the sales and the profits. Considering that, for the regarded period, the sales are high in the interval RF_1 from \$900 to \$1200, and the profits are acceptable in the interval RF_2, from 10 to 18%, select the best one of these companies, utilizing two different views: (a) sales and profit have equal weights in the decision; (b) profit has three times the weight of sales.

To solve, we must calculate the degrees of pertinence of the sales (Y_1) and of the profits (Y_2) of each company to the subsets RF_1 and RF_2, respectively, using the following membership functions:

Sales (X_1): $Y_1 = 0$, for $X_1 < 900$; $Y_1 = (X_1 - 900)/300$, for $900 \leq X_1 < 1200$; and $Y_1 = 1$, for $X_1 \geq 1200$;
Profit (X_2): $Y_2 = 0$, for $X_2 < 10\%$; $Y_2 = (X_2 - 10)/8$, for $10\% \leq X_2 < 18\%$; and $Y_2 = 1$, to $X_2 > 18\%$.

Then, we must calculate: (a) the arithmetic average or (b) the weighted average of the obtained membership degree and compare the value of the obtained averages for the eight companies. The one that presents the highest average will be considered the best company.

Table 8.1 summarizes the calculations that give the solution of the question.

It is concluded, then, that in the condition (a), H is the best company and, in the status (b), the best company is E.

It may occur that the average of the degrees of the pertinence of a company is maximum, despite one (or more) degrees of pertinence being zero, that is, despite the company being outside the acceptable conditions. In that case, an indicated (and necessary) verification for this process is calculating the products of the degrees of the pertinence of the factors. If this product results in 0 (zero), we will know that, at least about one of the factors, the company is outside the acceptable conditions and cannot be classified as the best one, even if it has the highest average.

Table 8.1 Calculation table for the selection of the best company

Firm	Sales (Factor F_1)	Profit (Factor F_2)	Membership to R_1 (Y_1)	Membership to $R_2(Y_2)$	Average (a) Y_{Ra}	Average (b) Y_{Rb}	Product $Y_1 \times Y_2$
A	750	7	0.00	0.00	0.00	0.00	0.00
B	600	14	0.00	0.50	0.25	0.38	0.00
C	800	17	0.00	0.88	0.44	0.66	0.00
D	850	12	0.00	0.25	0.13	0.19	0.00
E	990	18	0.30	1.00	0.65	0.83	0.30
F	1000	15	0.33	0.63	0.48	0.55	0.21
G	1100	14	0.67	0.50	0.58	0.54	0.33
H	1200	13	1.00	0.38	0.69	0.53	0.38

8.1.4 Another Application of the SVFDM (Shimizu [94], p. 65)

A company XYZ analyzes three alternatives (A_1 A_2 and A_3) of purchase, with spot payment of a property for the installation of a branch, based on three indicators (factors): net gain (F_1), distance from the property to the commercial center (F_2) and the total available area (F_3). F_2 and F_3 represent competitive advantages (Table 8.2).

Verify which is the best alternative in the cases when: (a) the three factors have equal weights; (b) the weights of the factors are 10, 5 and 8, respectively.

The first step for the solution is creating a membership function $Y = f(X)$ (which may be linear) for the favorable condition of each factor (F_i), imposing that the least favorable value of the factor corresponds to the membership degree 0 (zero) and the most favorable one, to 1 (one). That is if the growth of the factor favors the alternative, the membership degree 0 (zero) corresponds to the lowest value of the factor (increasing function); otherwise, 0 (zero) corresponds to the highest value (decreasing function).

The second step is calculating the membership degree of the factors for each one of the alternatives. Two of them, 0 (zero) and 1 (one), are already pre-established.

The third and last step is calculating the average of the membership degrees for each one of the alternatives and conclude for the best one, which is the one with the highest average. As seen in Table 8.3, (a) when the factors have equal weights, the

Table 8.2 Values of the indicators for the three alternatives

Factor		A_1	A_2	A_3
Net gain	F_1	470	500	420
Distance to shopping center	F_2	150	250	500
Total available area	F_3	600	400	1500

Table 8.3 Solution for a problem by SVFDM

1	2	3	4	5	6	7	8	9	10
Factor	Weight	A1 ($)	A2 (m)	A3 (m²)	Membership function		Membership degree		
					Monotonicity	Equation	A1	A2	A3
F1	10	470	500	420	increasing	$Y = (X - 420)/80$	0.63	1	0
F2	5	150	250	500	decreasing	$Y = (500 - X)/350$	1	0.71	0
F3	8	600	400	1500	increasing	$Y = (X - 400)/1100$	0.18	0	1
Sum	23				Average with equal weights		0.60	0.57	0.33
					Average with different weights		0.55	0.59	0.35

Table 8.4 Solution with presetting of the maximum and minimum values of the factors

1	2	3	4	5	6	7	8	9	10	11
Factor	Weight	A1 ($)	A2 (m)	A3 (m²)	Membership function			Membership degree		
					Monotonicity	Interval	Equation	A1	A2	A3
F1	10	470	500	420	increasing	300– 600	$Y = (X - 300)/300$	0.57	0.67	0.40
F2	5	150	250	500	decreasing	100– 600	$Y = (600 - X)/500$	0.90	0.70	0.20
F3	8	600	400	1500	increasing	400– 2000	$Y = (X - 400)/1600$	0.13	0,00	0.69
Sum	23						Average with equal weights	0.53	0.46	0.43
							Average with different weights	0.49	0.44	0.46

answer is alternative A_1; and (b) when the factors have the specified weights, the answer is A_2.

This exercise enables a more general view: assume that a company XYZ has specified in advance an interval with the maximum and minimum acceptable limits for each one of the factors, within its future interests (see column 7 of Table 8.4).

In this case, the degrees of pertinence, minimum (0) and maximum (1), would now correspond to these values, altering the membership functions.

Table 8.4 shows how the solution for this new situation would be, in which alternative A_1 was the best one, in both cases.

It must be observed that, in these cases, the membership function plays the role of what is commonly called utility function.

8.2   A More Elaborated Example for the Comparison of the Two Methods

To exemplify the two decision procedures, the feasibility analysis of the following enterprise will be studied by both processes: the launch of a new product in the market.

Several factors influence the success (or failure) of a new product launched in the market. Normally, the most influential factors to conduct the analysis are chosen. However, as it is an example, only the following factors will be considered:

F_1: **need and usefulness of the product**—Translated by the percentage (X_1) of the population (X) for which the product is essential in everyday life.

F_2: **acceptance of the product or similar product already existent in the market** —Translated by the percentage (X_2) of the population that utilizes it.

F_3: **product price in the market**—Translated by the ratio $(X_3$, in %) between the average product price (or of similar products) already existent in the market and its launch price.

F_4: **estimated product price**—Translated by the ratio $(X_4$, in %) between its cost and the average product price (or of similar products) in the market.

F_5: **time for development and implementation of the project and product launch**—Measured by the ratio $(X_5$, in %) between this time and the foreseen life cycle for the product.

F_6: **investment for development and implementation of the project and product launch**—Measured by the ratio $(X_6$, in %) between this investment and the expected net result in the foreseen life cycle for the product.

The issue we propose is to study the feasibility of launch of a new product in the market by the analysis of the conditions of the six chosen factors, obtained using a field research. Let's assume that this research led to the results X_i summarized by Table 8.5.

8.2.1   Solution by the Paraconsistent Decision Method—PDM

(a) **Establish the requirement level**. For this illustrative example, we adopted the value 0.50 as requirement level. It means that the analysis will be conclusive if the degree of certainty of the center of gravity is, in module, higher or equal to

Table 8.5 Table of the results obtained in the field research

F_i	F_1	F_2	F_3	F_4	F_5	F_6
X (%)	88	95	128	83	15	24

0.50 ($|H_W| \geq 0.50$), that is, if the degrees of evidence, favorable and contrary, final (of the center of gravity W) differ in, at least, 0.50. The decision will be favorable if the favorable evidence is, at least, 0.50 higher than the contrary evidence; and it will be unfavorable if the contrary evidence is, at least, 0.50 higher than the favorable one. With this requirement level, the decision rule is:

$H \geq 0.50 \Rightarrow$ favorable decision (feasible enterprise);
$H \leq -0.50 \Rightarrow$ unfavorable decision (unfeasible enterprise);
$-0.50 < H < 0.50 \Rightarrow$ inconclusive analysis.

(b) **Choose the most influential factors in the enterprise**. This has already been conducted at the beginning of Sect. 8.3.
(c) **Establish, for each factor, the sections** (S_j) that characterize the conditions in which every factor may be: S_1—favorable; S_2—indifferent; and S_3—unfavorable to the enterprise.

Factor F_1: S_1—above 70%; S_2—between 30 and 70%; and S_3—below 30%.
Factor F_2: S_1—above 70%; S_2—between 30 and 70%; and S_3—below 30%.
Factor F_3: S_1—above 110%; S_2—between 90 and 110%; and S_3—below 90%.
Factor F_4: S_1—below 50%; S_2—between 50 and 70%; and S_3—above 70%.
Factor F_5: S_1—below 30%; S_2—between 30 and 60%; and S_3—above 60%.
Factor F_6: S_1—below 40%; S_2—between 40 and 60%; and S_3—above 60%.

(d) **Build the database**, constituted by the degrees of favorable evidence (or belief) ($a_{i,j,k}$) and contrary evidence (or disbelief) ($b_{i,j,k}$) that each specialist (E_k) attributes to the success of the enterprise in the face of each influence factor (F_i), inside the conditions fixed by each one of the established sections (S_j). It will be assumed that four specialists have been chosen and that the degrees of evidence attributed by them are the ones in Table 8.6, which is the database.
(e) **Conduct the field research**. According to the conducted research, whose results are in Table 8.5, and which was established in 8.2.1, item c, the sections that translate the real conditions of the factors are (Table 8.7).
(f) **Apply the maximization (MAX) and minimization (MIN) rules** of logic Eτ to obtain the resultant degrees of favorable evidence ($a_{i,R}$) and contrary evidence ($b_{i,R}$) for each one of the factors.

For the application of the rules above, the experts must be divided into groups. It was assumed that they constituted two groups: Group A, with E_1 and E_2, and Group B, with E_3 and E_4. This way, the application of the rules is structured as follows:

Table 8.6 Database

Factor	Section	F and S	Special. 1		Special. 2		Special. 3		Special. 4	
			$a_{i,j,1}$	$b_{i,j,1}$	$a_{i,j,2}$	$b_{i,j,2}$	$a_{i,j,3}$	$b_{i,j,3}$	$a_{i,j,4}$	$b_{i,j,4}$
F_1	S_1	F1S1	0.88	0.04	0.94	0.14	0.84	0.08	0.78	0.03
	S_2	F1S2	0.48	0.43	0.53	0.44	0.58	0.39	0.48	0.41
	S_3	F1S3	0.01	0.94	0.13	0.88	0.14	1.00	0.17	0.91
F_2	S_1	F2S1	1.00	0.05	0.95	0.15	1.00	0.10	0.85	0.00
	S_2	F2S2	0.55	0.45	0.55	0.45	0.65	0.40	0.45	0.55
	S_3	F2S3	0.00	0.95	0.15	0.75	0.15	0.85	0.25	1.00
F_3	S_1	F3S1	0.92	0.08	0.98	0.18	0.88	0.12	0.82	0.07
	S_2	F3S2	0.52	0.47	0.57	0.48	0.62	0.43	0.52	0.45
	S_3	F3S3	0.05	0.98	0.17	0.83	0.18	0.02	0.21	0.95
F_4	S_1	F4S1	0.95	0.11	1.00	0.21	0.91	0.15	0.85	0.10
	S_2	F4S2	0.55	0.50	0.60	0.51	0.65	0.46	0.55	0.48
	S_3	F4S3	0.08	1.00	0.20	0.86	0.21	0.05	0.24	0.98
F_5	S_1	F5S1	1.00	0.88	0.90	0.00	0.95	0.15	0.94	0.05
	S_2	F5S2	0.50	0.50	0.60	0.50	0.60	0.40	0.50	0.40
	S_3	F5S3	0.00	1.00	0.10	0.80	0.90	0.08	1.00	0.15
F_6	S_1	F6S1	0.90	0.10	1.00	0.10	0.90	0.00	1.00	0.00
	S_2	F6S2	0.60	0.50	0.60	0.40	0.50	0.40	0.50	0.50
	S_3	F6S3	0.10	0.80	0.20	0.90	0.13	1.00	0.00	1.00

Table 8.7 Sections obtained in the field research

Factor F_i	F_1	F_2	F_3	F_4	F_5	F_6
X_i (%)	88	95	128	83	15	24
Section S_j	S_1	S_1	S_1	S_3	S_1	S_1

$$\text{MIN}\{\text{MAX}[E_1, E_2]; \text{MAX}[E_3, E_4]\} \quad \text{or}$$
$$\text{MIN}\{\text{Group A}; \text{Group B}\}$$

The applications of the rules are made with the aid of the PDM calculation table (Table 8.8). The sections S. obtained in the research are taken to column 2. The program seeks the corresponding opinions of the experts (column 3–10) in the database; applies the maximization rules inside the groups (intragroup) (columns 11–14) and minimization rules among the groups (intergroup) and obtains the resultant degrees of favorable evidence ($a_{i,R}$) and contrary evidence ($b_{i,R}$) for each one of the factors (columns 15 and 16); calculates the degrees of certainty (H_i) and uncertainty (G_i) for each factor (columns 17 and 18); and applies the decision rule, showing how each factor influences the enterprise (column 19).

Table 8.8 Calculation table of the Paraconsistent Decision Method (PDM)

1	2	3	4	5	6	7	8	9	10	11	12	13	14	15	16	17	18	19
Factor		Group A				Group B				Group A		Group B		MIN {A, B}		L of R = 0.50		
Section		Spec. 1		Spec. 2		Spec. 3		Spec. 4		MAX [E_1, E_2]		MAX [E_3, E_4]				Conclusions		
F_i	S_j	$a_{i,j,1}$	$b_{i,j,1}$	$a_{i,j,2}$	$b_{i,j,2}$	$a_{i,j,3}$	$b_{i,j,3}$	$a_{i,j,4}$	$b_{i,j,4}$	$a_{i,gA}$	$b_{i,gA}$	$a_{i,gB}$	$b_{i,gB}$	$a_{i,R}$	$b_{i,R}$	H	G	Decision
F_1	S_1	0.88	0.04	0.94	0.14	0.84	0.08	0.78	0.03	0.94	0.04	0.84	0.03	0.84	0.04	0.80	-0.12	VIABLE
F_2	S_1	1.00	0.05	0.95	0.15	1.00	0.10	0.85	0.00	1.00	0.05	1.00	0.00	1.00	0.05	0.95	0.05	VIABLE
F_3	S_1	0.92	0.08	0.98	0.18	0.88	0.12	0.82	0.07	0.98	0.08	0.88	0.07	0.88	0.08	0.80	-0.04	VIABLE
F_4	S_3	0.08	1.00	0.20	0.86	0.21	0.05	0.24	0.98	0.20	0.86	0.24	0.05	0.20	0.86	-0.66	0.06	UNVIABLE
F_5	S_1	1.00	0.88	0.90	0.00	0.95	0.15	0.94	0.05	1.00	0.00	0.95	0.05	0.95	0.05	0.90	0.00	VIABLE
F_6	S_1	0.90	0.10	1.00	0.10	0.90	0.00	1.00	0.00	1.00	0.10	1.00	0.00	1.00	0.10	0.90	0.10	VIABLE
Baricenter: average of the favorable and contrary degrees of evidence														**0.812**	**0.197**	**0.615**	**0.008**	VIABLE

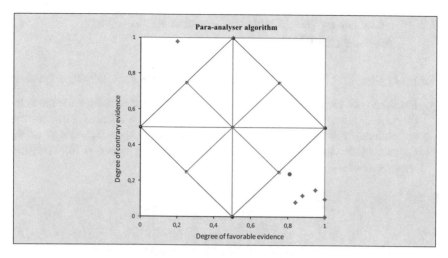

Fig. 8.5 Analysis of the result by the para-analyzer algorithm

(g) **Calculate the degrees of favorable evidence (a_w) and contrary evidence (b_w) of the center of gravity** of the points that represent the factors in the lattice of the annotations (para-analyzer algorithm). These translate the joint influence of the factors on the enterprise, and enable the final decision. a_W and b_W (last line of columns 15 and 16) are equal to the averages of the resultant degrees ($a_{i,R}$) and ($b_{i,R}$) obtained for each one of the factors. Besides that, the program calculates the degrees of certainty (H_W) and uncertainty (G_W) of the center of gravity (last line of columns 17 and 18), enabling the final decision making: **the enterprise is feasible** at the requirement level 0.50 (last line of column 19).

(h) **Make the decision.** As seen above, the program makes the decision, based on the decision rule. In fact, $H_W = 0.615$ and $0.615 \geq 0.50$ implicate in $H_W \geq NE$, which implicates in a favorable decision, that is, the enterprise is feasible.

The analysis of the result may also be performed by the para-analyzer algorithm, as shown in Fig. 8.5.

The analysis of the chart shows that five factors belong to the truth region, being, then, favorable to the enterprise; one of them belongs to the falsity region, being contrary to the company. As the center of gravity is in the truth region, it is concluded, by the analysis of these six factors in the researched conditions, that the enterprise is feasible.

8.2.2   Solution by the Simplified Version of the Fuzzy Method (SVFDM)

For the application of the simplified version, we execute the following sequence.

(a) **Establish the requirement level**. Analogously to what was done for the PDM (Sect. 8.2.1, item a), the requirement level 0.50 will be adopted. It means that the analysis will be conclusive if the module of the average ($|\mathbf{Y_R}|$) of the degrees of the pertinence of the factors is equal or above 0.50. With this requirement level, the decision rule is:

> $\mathbf{Y_R} \geq 0.50 \Rightarrow$ **favorable decision (feasible enterprise)**;
> $\mathbf{Y_R} \leq -0.50 \Rightarrow$ **unfavorable decision (unfeasible enterprise)**;
> $-0.50 < YR < 0.50 \Rightarrow$ **inconclusive analysis**.

(b) **Choose the most influential factors in the enterprise**. This has already been conducted at the beginning of Sect. 8.2.

(c) **Establish, for each factor, the ranges that translate the conditions considered as favorable (RF) and as contrary (RC) to the enterprise, and determine the pertinence function for each one**.

For each one of the factors F_i we have a value of the membership degree (Y_i) to the unfavorable condition or the favorable condition, according to the real condition of the factor (X_i). The interval of X_i for which the condition is considered favorable (or unfavorable) is defined by a specialist (or specialists) in the area.

The conditions, favorable (RF) or contrary (RC), are translated by the intervals inside which the adapted membership degree (Y_i) varies from 0 to 1, or from 0 to -1, respectively. For the six chosen factors, the established conditions and the membership functions are the ones presented below.

F_1: **need and usefulness of the product**—Translated by the percentage of the population (X_1, in %) for which the product is essential in everyday life.

RC: between 10 and 30%; RF: between 70 and 90%.
Results in: $Y = 0.05 (X - 30)$, for $10 \leq X \leq 30$; $Y = 0.05 (X - 70)$, for $70 \leq X \leq 90$; $Y = 0$, for $30 \leq X \leq 70$; $Y = -1$, for $X \leq 10$, and $Y = 1$, for $X \geq 90$.

For the factor F_1, the membership degree Y varies with the variable X (in %) according to the chart of Fig. 8.6.

F_2: **acceptance of the product or similar product already existent in the market**—Translated by the percentage (X_2, in %) of the population that utilizes it.

RC: between 10 and 30%; RF: between 70 and 90%.

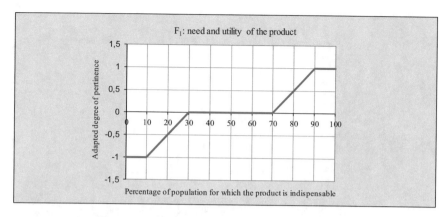

Fig. 8.6 Variation of the membership degree for factor F_1

Results in: $Y = 0.05 \ (X - 30)$, for $10 \leq X \leq 30$; $Y = 0.05 \ (X - 70)$, for $70 \leq X \leq 90$; $Y = 0$, for $30 \leq X \leq 70$; $Y = -1$, for $X \leq 10$, and $Y = 1$, for $X \geq 90$.

F_3: product price in the market—Translated by the ratio (X_3, in %) between the average product price (or of similar products) already existent in the market and its launch price.

RC: between 50 and 90%; RF: between 110 and 150%.
Results in: $Y = 0.025 \ (X - 90)$, for $50 \leq X \leq 90$; $Y = 0.025 \ (X - 110)$, for $110 \leq X \leq 150$; $Y = 0$, for $90 \leq X \leq 110$; $Y = -1$, for $X \leq 50$, and $Y = 1$, for $X \geq 150$.

F_4: estimated product cost—Translated by the ratio (X_4, in %) between its cost and the average product price (or of similar products) in the market.

RC: between 70 and 90%; RF: between 30 and 50% (Fig. 8.7)
Results in: $Y = 0.05 \ (70 - X)$, for $70 \leq X \leq 90$; $Y = 0.05 \ (50 - X)$, for $30 \leq X \leq 50$; $Y = 0$, for $50 \leq X \leq 70$; $Y = -1$, for $X \geq 90$, and $Y = 1$, for $X \leq 30$.

F_5: time for development and implementation of the project and product launch—Measured by the ratio (X_5, in %) between this time and the foreseen life cycle for the product.

RC: between 60 and 70%; RF: between 10 and 30%.
Results in: $Y = 0.10 \ (60 - X)$, for $60 \leq X \leq 70$; $Y = 0.05 \ (30 - X)$, for $10 \leq X \leq 30$; $Y = 0$, for $30 \leq X \leq 60$; $Y = -1$, for $X \geq 70$, and $Y = 1$, for $X \leq 10$.

F_6: investment for development and implementation of the project and product launch—Measured by the ratio (X_6, in %) between this investment and the expected net result in the foreseen life cycle for the product.

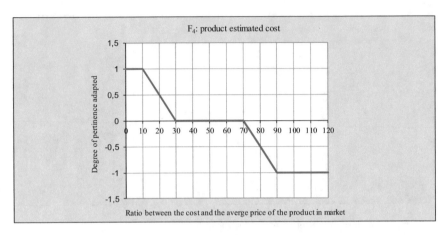

Fig. 8.7 Variation of the membership degree for factor F_4

Table 8.9 Table of the results obtained in the field research

F_i	F_1	F_2	F_3	F_4	F_5	F_6
X (%)	88	95	128	83	15	24

Table 8.10 Calculation and decision table

Factor	Condition of the factor	Membership degree	Decision
F_i	X_i	Y_i	0.50
F1	88	0.90	VIABLE
F2	95	1.00	VIABLE
F3	128	0.45	**NOT CONCLUSIVE**
F4	83	−0.65	UNVIABLE
F5	15	0.75	VIABLE
F6	24	0.80	VIABLE
	YR =	**0.54**	VIABLE

RC: between 60 and 80%; RF: between 20 and 40%.
Results in: $Y = 0.05\ (60 - X)$, for $60 \leq X \leq 80$; $Y = 0.05\ (40 - X)$, for $20 \leq X \leq 40$; $Y = 0$, for $40 \leq X \leq 60$; $Y = -1$, for $X \geq 80$, and $Y = 1$, for $X \leq 20$.

(d) **Conduct the field research**. The research has been performed and the results are in Table 8.9, repeated here.

These values of X_i are placed in column 2 of the calculation table (Table 8.10), which displays the corresponding Y values, (column 3). Using the membership functions, it calculates Y_R (last line of column 3) and makes the decision (column 4).

By the result obtained in Table 8.9, it is observed that four factors (F_1, F_2, F_5 and F_6) are in favorable conditions to the enterprise; F_4 in unfavorable condition and F_3 is indifferent, everything at the requirement level of NE = 0.50. The joint influence of the six factors shows that the enterprise is feasible at this requirement level.

8.3   Comparison Between the Two Methods

The decision by the PDM is based on the resultant degree of certainty for the center of gravity (H_W), which varies from −1 to 1. Analogously, the decision by the SVFDM is based on the resultant membership degree (Y_R) adapted to the subsets that define the conditions, favorable (RF) or contrary (RC), calculated according to the variable that identifies the factor. This degree also varies from −1 to 1.

When $H_W > 0$, we have a favorable condition to the enterprise, but very weak, as it means that only the degree of favorable evidence is higher than the degree of contrary evidence, as the regions of $H_W > 0$ and $H_W < 0$ are adjoining. That is the reason for the convenience of establishing the requirement level relatively greater than zero.

On the other hand, when $Y_R > 0$, we already have a stronger favorable condition to the enterprise, as there is an interval in which Y = 0, relatively high, separating the region of Y > 0 (favorable) from the region of Y < 0 (unfavorable). Therefore, in this case, just the fact of the requirement level being greater than zero is sufficient for us to have a right decision.

Observe in Fig. 8.8 that, in identical conditions, for $Y_R > 0$, we have in correspondence $H_W > 0.50$ and, for $Y_R > 0.50$, we have $H_W > 0.75$.

Given that, we conclude then, when establishing the same value for the requirement level, the fuzzy decision is stronger than the paraconsistent one, as observed in Fig. 8.8. In the represented case, the value NE = 0.5 for the fuzzy corresponds to the value NE = 0.75 for the paraconsistent decision.

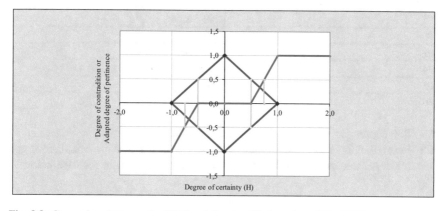

Fig. 8.8 Comparison between the PDM and the simplified version of the SDM

Finally, we remind you that the whole analysis is surrounded by subjectivity, once the degrees of evidence (in the PDM) and the ranges (in the SVFDM) are established ny experts.

Exercises

8.1 Suppose that a businessman has to decide among five companies he intends to acquire: Z_1 to Z_5. He has selected six indicators (factors) that influence the value of a company, F_1 to F_6, such as annual gross revenue, profit percentage, growth perspective, etc. Considering the table below, which provides the values of the indicators for the five companies, determine which is the most indicated one.

Factor	Z_1	Z_2	Z_3	Z_4	Z_5
F1	200	280	400	520	600
F2	27	20	32	45	38
F3	200	75	125	180	110
F4	15	35	42	65	42
F5	300	200	125	275	225
F6	12	8	15	20	10

For factors F_2 and F_5, the competitive edge reduces when the value increases.

8.2 In relation to exercise 8.1, which would the most indicated company be, if the weights below were adopted for the factors?

Factor	F1	F2	F3	F4	F5	F6
Weight	2	4	6	1	6	3

8.3 Solve exercise 8.1, assuming that the intervals of desirable values for each factor are:

Factor	Interval
F1	250–500
F2	20–40
F3	100–300
F4	10–60
F5	100–300
F6	5–25

8.4 Solve exercise 8.2, using the condition of exercise 8.3.

8.5 Using the data base BD-E.g. 8.5 of Appendix C, perform the calculations and make the decision, applying the PDM and the SVFDM, in the case below.

The sections that determine the conditions, favorable (S_1) and unfavorable (S_3) to the success of an enterprise, in which the six factors are found, are:

$$(\text{Factor } F_1)--S_1 : 120 \leq X \leq 160 \quad \text{and} \quad S_3 : 40 \leq X \leq 80;$$
$$(\text{Factor } F_2)--S_1 : 10 \leq X \leq 30 \quad \text{and} \quad S_3 : 70 \leq X \leq 100;$$
$$(\text{Factor } F_3)--S_1 : 30 \leq X \leq 80 \quad \text{and} \quad S_3 : 120 \leq X \leq 170;$$
$$(\text{Factor } F_4)--S_1 : 25 \leq X \leq 45 \quad \text{and} \quad S_3 : 0 \leq X \leq 15;$$
$$(\text{Factor } F_5)--S_1 : 250 \leq X \leq 500 \quad \text{and} \quad S_3 : 50 \leq X \leq 200;$$
$$(\text{Factor } F_6)--S_1 : 100 \leq X \leq 500 \quad \text{and} \quad S_3 : 800 \leq X \leq 1200;$$

The values in between these intervals characterize the indifferent situation (S_2); above or below these intervals, they follow the status of the closest interval. The X values that characterize the factors in the field research are the following:

Factor	F1	F2	F3	F4	F5	F6
X	150	80	25	20	450	400

To apply the PDM, consider three groups: A, constituted by experts E_1 and E_2; B, by E_3 and C, composed by E_4. Adopt the requirement level NE = 0.50.

8.6 Solve exercise 8.5, considering all the factors with equal weights.

8.7 Solve exercise 8.5, considering that the X values that characterize the factors in the field research are the following:

Factor	F1	F2	F3	F4	F5	F6
X	50	80	160	30	30	700

8.8 Solve exercise 8.7, considering all the factors with equal weights.

Answers

8.1 The averages of the membership degrees for each company are:

Z_1	Z_2	Z_3	Z_4	Z_5
0.34	0.36	0.59	0.63	0.45

Therefore, answer Z_4.

8.2 The averages of the membership degrees for each company are:

Z_1	Z_2	Z_3	Z_4	Z_5
0.45	0.37	0.63	0.52	0.38

Therefore, answer Z_3.

8.3 The averages of the membership degrees for each company are:

Z_1	Z_2	Z_3	Z_4	Z_5
0.27	0.38	0.52	*0.55*	0.40

Therefore, answer Z_4.

8.4 The averages of the membership degrees for each company are:

Z_1	Z_2	Z_3	Z_4	Z_5
0.31	0.37	*0.50*	0.38	0.29

Therefore, answer Z_3.

8.5 By the PDM: $W = (0.78; 0.22)$; $H_w = 0.56$; Feasible enterprise. By the SVFDM: $Y_R = 0.66$; Feasible enterprise.

8.6 By the PDM: $W = (0.65; 0.35)$; $H_w = 0.30$; Inconclusive analysis. By the SVFDM: $Y_R = 0.41$; Inconclusive analysis.

8.7 By the PDM: $W = (0.13; 0.80)$; $H_w = -0.68$; Infeasible enterprise. By the SVFDM: $Y_R = -0.65$; Infeasible enterprise.

8.8 By the PDM: $W = (0.23; 0.72)$; $H_w = -0.48$; Inconclusive analysis. By the SVFDM: $Y_R = -0.44$; Inconclusive analysis.

Chapter 9
Complementary Reading: An Example from Everyday Life

As seen in the Preface, the maximization rule is applied to the ***maximize of the degree of favorable evidence*** inside each group (**MAX** operator) and the ***minimizing of the degree of contrary evidence*** among the groups (**MIN** operator). Thus such evidence is achieved, first, maximizing the degrees of favorable evidence and minimizing the degrees of contrary evidence inside the groups and, then, minimizing the degrees of favorable evidence and maximizing the degrees of contrary evidence among the groups, utilizing the results obtained by the application of the first.

However, in the certain analysis it may be applied to ***maximize the values of the degrees of evidence***, favorable and contrary, inside each group (**OR** operator). The minimization rule may be implemented among the groups to ***minimize the maximum values of the degrees of evidence***, favorable and contrary, obtained by the application of the first inside each group (**AND** operator). This interpretation has the advantage of taking the most predictable and coherent result, but on the other hand, it has the disadvantage of not capturing the contradictions of the database so easily.

To enable the reader to make the comparison between the two ways to apply the maximization and minimization rules, the example below will be analyzed in detail. For those who still have not captured the criterion for the formation of the groups for the application of the Logic Eτ rules, we believe this example will make the idea clearer.

Imagine the four sectors of a soccer team: **A**—the goalkeeper (a player with number 1), **B**—the defense (four players numbered from 2 to 5); **C**—the midfield (three players numbered from 6 to 8) and **D**—the attack (three players numbered from 9 to 11). That is what the soccer players call 4-3-3 scheme.

Each player, each team sector or, also, the whole team may be classified in categories according to the following descending ordinal scale: **Great, Good, Medium, Regular** and **Weak**. To each player, a degree of favorable evidence (*a*) and a degree of contrary evidence (*b*) may be attributed, which translates the expectation of its performance as a result of its past actions. This way, a great player

© Springer International Publishing AG, part of Springer Nature 2018
F. R. de Carvalho and J. M. Abe, *A Paraconsistent Decision-Making Method*,
Smart Innovation, Systems and Technologies 87,
https://doi.org/10.1007/978-3-319-74110-9_9

Table 9.1 Categories adopted for the classification of the players, of the sectors and the team

Category	Degree of favorable evidence (a)	Degree of contrary evidence (b)	Degree of certainty (H)
Optimum	$0.8 \leq a \leq 1.0$	$0.0 \leq b \leq 0.2$	$0.6 \leq H \leq 1.0$
Good	$0.6 \leq a < 0.8$	$0.2 < b \leq 0.4$	$0.2 \leq H < 0.6$
Medium	$0.4 \leq a < 0.6$	$0.4 < b \leq 0.6$	$-0.2 \leq H < 0.2$
Regular	$0.2 \leq a < 0.4$	$0.6 < b \leq 0.8$	$-0.6 \leq H < -0.2$
Weak	$0.0 \leq a < 0.2$	$0.8 < b \leq 1.0$	$-1.0 \leq H < -0.6$

is characterized by a high degree of favorable evidence (of belief) and low degree of contrary evidence (of disbelief).

Although it is not necessary, to simplify the considerations of this example, it will be assumed that the degrees of favorable evidence and contrary evidence are Boolean complements, that is, $a + b = 1$. With that, we assume that the data referring to each player (the annotations) do not present contradiction. In fact, with this hypothesis, the degree of the contradiction of each player (which is defined by the expression $G(a; b) = a + b - 1$) is always null.

Besides that, a player will be considered "Great" if se $0.8 \leq a \leq 1.0$ and $0.0 \leq b \leq 0.2$; "Good", if $0.6 \leq a < 0.8$ and $0.2 < b \leq 0.4$; "Medium", if $0.4 \leq a < 0.6$ and $0.4 < b \leq 0.6$; "Regular", if $0.2 \leq a < 0.4$ and $0.6 < b$ 0.8; and "Weak", if $0.0 \leq a < 0.2$ and $0.8 < b < 1.0$.

As a consequence, being $H = a - b$ the degree of certainty, it will be said that the player is "Great" if $0.6 \leq H \leq 1.0$; "Good", if $0.2 \leq H \leq 0.6$; "Medium", if $-0.2 \leq H < 0.2$; "Regular", if $-0.6 \leq H < -0.2$; and "Weak", if $-1.0 \leq H < -0.6$. By analogy and coherence, the same criterion will be utilized to classify each sector of the team and the whole team.

Table 9.1 summarizes the intervals that characterize the criteria to frame the players, the sectors of the team and the whole team in the five established categories.

A coach (here, the knowledge engineer) understands that the classification of any sector of the team is given by the classification of its best player. Thus, the sector in which the best player is "Good" will be classified as "Good"; the sector in which the best player is "Medium" will be classified as "Medium", etc. Of course, for the sector to be "Great", it must have, at least, one "Great" player. That is, the sector is classified by the player with maximum classification, which justifies applying the rule for maximization of the degree of certainty inside each group (of each sector of the team).

On the contrary, the coach understands that the classification of the team is given by the classification of its most deficient sector. Thus, if the team has the worst sector classified as "Good", it will be classified as "Good", regardless of the other sectors being classified as "Great"; if the worst sector is "Medium", the team will be classified as "Medium", regardless of the other sectors being classified as "Great" or as "Good", etc. Evidently, for the team to be "Great", all the sectors must be classified as "Great". Therefore, the team is classified by the minimum, that is, by the most deficient sector, which justifies the application of the rule for minimization of the degree of certainty among the groups (sectors of the team).

Repeating: for the sector to be "Medium", for example, it is sufficient that the best player of the sector is "Medium" (classified by the best, by the maximum). However, for the team to be "Medium", the best sector being "Medium" is not enough; on the contrary, it is required that the worst sector be "Medium" (classified by the worst, by the minimum).

Therefore, for a team to be "Great", it must have all the sectors classified as "Great", and for such purpose, each sector must have, at least, one "Great" player; for the team to be "Good", it must have all the sectors classified as, at least, "Good", and for that, each sector must have, at least, one at least "Good" player, etc.

The goalkeeper, as he is alone in the group, determines the maximum limit of classification of the team, that is, if the goalkeeper is "Good", the team may be, at most, "Good", regardless of all the other players; if he is "Medium", at most "Medium", etc. The same is valid for the best player of each sector.

Thus, in a feasibility analysis of the team, the groups are already naturally constituted. The goalkeeper, who is the only one in the sector, constitutes one group (A); the four defense players constitute another group (B), as just one of them being "Great" is sufficient to meet the requirement of a "Great" team; analogously, the three of the midfield constitute the third group (C) and the three attackers, the fourth group (D).

The distribution of the groups for the application of the **MAX** and **MIN** operators is the following:

$$\textbf{MIN}\{[Group\ A],\ [Group\ B],\ [Group\ C],\ [Group\ D]\}\ \text{or}$$
$$\textbf{MIN}\{[1],\textbf{MAX}\ [2,\ 3,\ 4,\ 5],\textbf{MAX}\ [6,\ 7,\ 8],\textbf{MAX}\ [9,\ 10,\ 11]\}\ \text{or}$$
$$\textbf{MIN}\{[(a_A;b_A)],\ [(a_B;b_B)],\ [(a_C;b_C)],\ [(a_D;b_D)]\}$$

which may be represented by the scheme of Fig. 9.1.

To test and compare the two interpretations given to the applications of the maximization and minimization rules, a database will be created, adopting the following criterion. As already said, each player will be attributed the pair of annotations $(a;b)$, in which $b = 1 - a$ (as it was assumed that $a + b = 1$).

The supremum of the interval that defines the degree of favorable evidence of each category (Table 9.1) will be called **S**. Thus, for the categories defined above, we have: "Great": $\mathbf{S} = 1.00$; "Good": $\mathbf{S} = 0.80$; "Medium": $\mathbf{S} = 0.60$; "Regular": $\mathbf{S} = 0.40$; and "Weak": $\mathbf{S} = 0.20$.

For player 1 (goalkeeper), the degree of favorable evidence (or degree of belief we have in his performance) will be defined by $a = \mathbf{S} - 0.16$.

Therefore, if the goalkeeper is "Great", his degree of favorable evidence is: $a = \mathbf{1.00} - 0.16 = 0.84$; if he is "Good", $a = \mathbf{0.80} - 0.16 = 0.64$; "Medium", $a = \mathbf{0.60} - 0.16 = 0.44$; "Regular", $a = \mathbf{0.40} - 0.16 = 0.24$; and if he is "Weak", $a = \mathbf{0.20} - 0.16 = 0.04$.

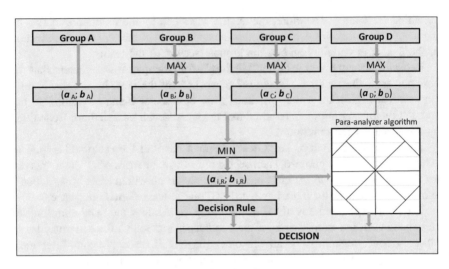

Fig. 9.1 Scheme for application of the **MAX** and **MIN** operators

With an analogous thought for the other sectors, it will be adopted: sector B, $a = S - 0.12$; sector C, $a = S - 0.08$; and sector D, $a = S - 0.04$. Table 9.2 summarizes the adopted values for a in the four groups.

Besides that, for sectors B, C and D, it will be assumed that the category of the player decreases one level when his number increases in 1 (one). This way, in group B, the best player is number 2. Therefore, if 2 is "Great", then 3 is "Good", 4 is "Medium" and 5 is "Regular"; if 2 is "Good", then 3 is "Medium", 4 is "Regular" and 5 is "Weak"; but, if 2 is "Medium", 3 is "Regular" and 4 is "Weak", what will the category of player 5 be? For these considerations, it will be assumed that number 5 is also "Weak". That is, whenever a player is "Weak", the successive one (s) of the same sector will also be considered "Weak".

With these criteria, degrees of favorable and contrary evidence may be attributed to all the players of the team in all the possible combinations. However, how many combinations are these? As there are 5 possible categories for each sector of the team and there are 4 sectors, the rule of product teaches us that there are 54 = 625 possibilities. In fact, there are 5 possibilities for sector A which, multiplied by 5 possibilities of sector B, by 5 of C and by 5 of sector D, resulting in $5 \times 5 \times 5 \times 5 = 5^4 = 625$.

Out of these 625 possibilities, in how many of them are the team "Great"? Just remember that, for the team to be "Great", all it sectors must be classified as "Great". As each sector only has 1 possibility of being "Great", once more the rule of product:

$$1 \times 1 \times 1 \times 1 = 1^4 = 1,$$

Table 9.2 Values of the degrees of favorable evidence (a)

Category	Degree of favorable evidence (a)	Supremum of the a interval S	Sector A $a = S - 0.16$	Sector B $a = S - 0.12$	Sector C $a = S - 0.08$	Sector D $a = S - 0.04$
Optimum	$0.8 \leq a \leq 1.0$	1.0	0.84	0.88	0.92	0.96
Good	$0.6 \leq a < 0.8$	0.8	0.64	0.68	0.72	0.76
Medium	$0.4 \leq a < 0.6$	0.6	0.44	0.48	0.52	0.56
Regular	$0.2 \leq a < 0.4$	0.4	0.24	0.28	0.32	0.36
Weak	$0.0 \leq a < 0.2$	0.2	0.04	0.08	0.12	0.16

that is, there is only one chance for the team to be "Great".

Moreover, how many are the possibilities of the team being "Good"? To be "Good", all its sectors must be classified as, at least, "Good", that is, each sector may have "Good" or "Great". Thus, there are two possibilities for each sector, which enables us to calculate:

$$2 \times 2 \times 2 \times 2 = 2^4 = 16$$

Then the team has 8 possibilities of being "Good". Right? No, because in this calculation, the previous case was included, in which all the sectors are classified as "Great".

Therefore, there are $16 - 1 = 2^4 - 1^4 = 15$ possibilities of the team being "Good".

Now, you calculate, reader: how many possibilities are there of the team being "Medium", "Regular" or "Weak"? The answers are at the end of this chapter.

The following task is to assemble a table with 625 lines (to analyze all the possibilities), initially, with 22 columns (two, *a* and *b*, for each one of the 11 players), from column A to V, following the rules for obtention of the values of *a* and *b*. Start in line 6, leaving the first five ones for headers.

For each player (1–11), calculate the degree of certainty and assemble an Excel formula to classify him in one of the five categories, in the 625 possibilities (hint: function SE of Excel). With that, you will be adding 22 more columns to the table (from W to AR).

After doing that, copy this page of the Excel spreadsheet on another page of the same spreadsheet.

On the first page, do the following:

(a) Calculate, applying the MAX operator, the degrees of evidence, favorable and contrary, for each one of the sectors (groups) of the team (columns from AS to AZ);

(b) Calculate the degree of certainty (H = a − b) of each sector of the team and, using the same formula from the previous paragraph, classify each one of the sectors (columns BA–BH);

(c) Applying the MIN operator, determine the resultant degrees of evidence, favorable and contrary, for the team, in each one of the 625 possibilities (columns BI and BJ);

(d) Calculate the degree of certainty ($H = a - b$) of the team and classify it, utilizing the same formula in item b, above (columns BK and BL); with these four items, 20 more columns were added to the table, reaching 64 columns;

(e) Find a formula (in cells BL 631–635) to calculate the quantity of times the team results in each one of the five categories: "Great", "Good", "Medium", "Regular" or "Weak" (hint: function **CONT.SE** of Excel). To check, do not forget that the sum of these quantities must be 625.

On the second page, repeat the sequence of the previous paragraph, but applying the **OR** and **AND** operators, instead of the **MAX** and **MIN** operators, respectively.

Compare the results and draw your conclusions.

With the same formula of item (e), calculate the quantity of times each player and each sector of the team results in each one of the five categories.

Answering to the previously done questioning, the possibilities of the team being "Medium", "Regular" or "Weak" are $3^4 - 2^4 = 65$, $4^4 - 3^4 = 175$ and $5^4 - 4^4 = 369$, respectively.

See which were these results obtained by the counts performed in the tables of the two pages of the spreadsheet: one by the application of the **MAX** and **MIN** operators and the other with the **OR** and **AND** operators (cells BL 631–635).

Compare the results and draw your conclusions.

Another verification that may be done is to analyze, for each category of the team, in which categories the sectors are. For this purpose, just apply a filter in the tables, filter, in column BL, the different categories of the team and verify in which categories the different sectors are. The result is more remarkable and interesting for the categories "Great" and "Good" of the team.

However, if you, reader, did not want to face this challenge of making two spreadsheets, each one with 625 lines and 64 columns, don't be upset, because of these spreadsheets, already elaborated, may be found in Appendix E.

Besides that, you will be able to assemble other databases utilizing other criteria, perform the tasks of the last paragraphs and draw new conclusions. Also, try to assemble a database without a and b being complementary, that is, without having $a + b = 1$. If you want to have less work, think of an indoor soccer team, which has five components: one goalkeeper, two in the defense and two in the attack.

Chapter 10
An Overview of More Applications

10.1 Introduction

Annotated logics are surprisingly useful logics. In what follows, we mention some other applications of the annotated logics (mainly the logic Eτ).

We begin with the paraconsistent logic programming language—ParaLog

In da Costa et al. [95], Abe et al. [99] it was developed a paraconsistent logic programming language—Paralog. As it is well known, the development of computationally efficient programs in it should exploit the following two aspects in its language:

1. The declarative aspect that describes the logic structure of the problem, and
2. The procedural aspect that describes how the computer solves the problem.

However, it is not always an easy task to conciliate both aspects. Therefore, programs to be implemented in Paralog should be well defined to evidence both the declarative aspect and the procedural side of the language.

It must be pointed out that programs in Paralog, like programs in standard Prolog, may be easily understood or reduced—when well defined—using addition or elimination of clauses, respectively.

A small knowledge base in the domain of Medicine is presented as a Paralog program. The development of this small knowledge base was subsidized by the information provided by three experts in Medicine. The first two specialists—clinicians—provided six[1] diagnosis rules for two diseases: disease1 and disease2. The last specialist—a pathologist—provided information on four symptoms: symptom1, symptom2, symptom3 and symptom4.

Example 1 A small knowledge base in Medicine implemented in Paralog

```
disease1(X): [1.0, 0.0] < -
```

---

[1]The first four diagnosis rules were supplied by the first expert clinician and the two remaining diagnosis rules were provided by the second expert clinician.

© Springer International Publishing AG, part of Springer Nature 2018
F. R. de Carvalho and J. M. Abe, *A Paraconsistent Decision-Making Method*,
Smart Innovation, Systems and Technologies 87,
https://doi.org/10.1007/978-3-319-74110-9_10

```
symptom1(X): [1.0, 0.0] &
symptom2(X): [1.0, 0.0]
disease2(X): [1.0, 0.0] < -
symptom1(X): [1.0, 0.0] &
symptom3(X): [1.0, 0.0]
disease1(X): [0.0, 1.0] < -
disease2(X): [1.0, 0.0].
disease2(X): [0.0, 1.0] < -
disease1(X): [1.0, 0.0].
disease1(X): [1.0, 0.0] < -
symptom1(X): [1.0, 0.0] &
symptom4(X): [1.0, 0.0].
disease2(X): [1.0, 0.0] < -
symptom1(X): [0.0, 1.0] &
symptom3(X): [1.0, 0.0].
symptom1(john): [1.0, 0.0].
symptom1(bill): [0.0, 1.0].
symptom2(john): [0.0, 1.0].
symptom2(bill): [0.0, 1.0].
symptom3(john): [1.0, 0.0].
symptom3(bill): [1.0, 0.0].
symptom4(john): [1.0, 0.0].
symptom4(bill): [0.0, 1.0].
```

In this example, several types of queries can be performed. Table 10.1 below shows some query types, the evidences provided as answers by the Paralog inference engine and their respective meaning.

The knowledge base implemented in the example may also be implemented in standard Prolog, as shown in the example.

Example 2 The knowledge base of Example 10.1 implemented in standard Prolog

```
disease1(X):-
symptom1(X),
symptom2(X).
disease2(X):-
symptom1(X),
symptom3(X).
disease1(X):-
not disease2(X).
disease2(X):-
not disease1(X).
disease1(X):-
symptom1(X),
symptom4(X).
disease2(X):-
```

Table 10.1 Query and answer forms in Paralog

Item	Query and answer form		Meaning
1	Query	Disease1(bill): [1.0, 0.0]	Does Bill have disease1?
	Evidence	[0.0, 0.0]	The information on Bill's disease1 is unknown
2	Query	Disease2(bill): [1.0, 0.0]	Does Bill have disease2?
	Evidence	[1.0, 0.0]	Bill has disease2
3	Query	Disease1(john): [1.0, 0.0]	Does John have disease1?
	Evidence	[1.0, 1.0]	The information on John's disease1 is inconsistent
4	Query	Disease2(john): [1.0, 0.0]	Does John have disease2?
	Evidence	[1.0, 1.0]	The information on John's disease2 is inconsistent
5	Query	Disease1(bob):[1.0, 0.0]	Does Bob have disease1?
	Evidence	[0.0, 0.0]	The information on Bob's disease1 is unknown

```
not symptom1(X),
symptom3(X).
symptom1(john).
symptom3(john).
symptom3(bill).
symptom4(john)
```

In this example, several types of queries can be performed as well. Table 10.3 shows some query types provided as answers by the standard Prolog and their respective meaning.

Starting from Examples 10.1 and 10.2 it can be seen that there are different characteristics between implementing and consulting in Paralog and standard Prolog. Among these features, the most important are:

1. The semantic characteristic; and
2. The execution control characteristic.

The first characteristic may be intuitively observed when the program codes in Examples 1 and 2 are placed side by side. That is, when compared to Paralog, the standard Prolog representation causes loss of semantic information on facts and rules. This is because standard Prolog cannot directly represent the negation of facts and rules.

In Example 10.1, Paralog program presents a four-valued evidence representation. However, the information loss may be greater for a standard Prolog program, if the facts and rules of Paralog use the intermediate evidence of lattice $\tau = \{x \in \Re|$ $0 \leq x \leq 1\} \times \{x \in \Re| 0 \leq x \leq 1\}$. This last characteristic may be observed in

Table 10.2 Query and answer forms in standard Prolog

Item	Query and answer form		Meaning
1	Query	Disease1(bill)	Does Bill have disease1?
	Answer	Loop	System enters into an infinite loop
2	Query	Disease2(bill)	Does Bill have disease2?
	Answer	Loop	System enters into an infinite loop
3	Query	Disease1(john)	Does John have disease1?
	Answer	Yes	John has disease1
4	Query	Disease2(john)	Does John have disease2?
	Answer	Yes	John has disease2
5	Query	Disease1(bob)	Does Bob have disease1?
	Answer	No	Bob does not have disease1

the Table 10.2. The Table 10.2 shows five queries and answers, presented and obtained both in Paralog and standard Prolog program.

The answers obtained from the two approaches present major differences. That is, to the first query: "Does Bill have disease1?", Paralog answers that the information on Bill's disease1 is unknown, while the standard Prolog enters into a loop. This happens because the standard Prolog inference engine depends on the ordination of facts and rules to reach deductions. This, for standard Prolog to be able to deduct an answer similar to Paralog, the facts and rules in Example 10.2 should be reordered. On the other hand, as the Paralog inference engine does not depend on reordering facts and rules, such reordering becomes unnecessary.

In the second query: "Does Bill have disease2?", Paralog answers that "Bill has disease2", while the standard Prolog enters into a loop. This happens for the same reasons explained in the foregoing item.

In the third query: "Does John have disease1?", Paralog answers that the information on John's disease1 is inconsistent, while the standard Prolog answers that "John has disease1". This happens because the standard Prolog inference engine, after reaching a conclusion that "John has disease1" does not check whether there are other conclusions leading to a contraction. On the other hand, Paralog performs such check, leading to more appropriate conclusions.

In the fourth query: "Does John have disease2?", Paralog answers that the information on John's disease2 is inconsistent while the standard Prolog answers that "John has disease2". This happens for the same reasons explained in the preceding item.

In the last query: "Does Bob have disease1", Paralog e answers that the information on Bob's disease1 is unknown, while the standard Prolog answers that "Bob does not have disease1". This happens because the standard Prolog inference engine does not distinguish the two possible interpretations for the answer not. On the other hand, the Paralog inference engine, being based on an infinitely valued paraconsistent evidential logic, allows the distinction to be made.

In view of the above, it is shown that the use of the Paralog language may handle several Computer Science questions more naturally.

10.2   Automation and Robotics

The Paracontrol is the eletric-eletronic materialization of the Para-analyzer algorithm Da Silva Filho [99], which is basically an electronic circuit, which treats logical signals in a context of logic Eτ. Such circuit compares logical values and determines domains of a state lattice corresponding to the output value. Favorable evidence and contrary evidence degrees are represented by a voltage. Certainty and Uncertainty degrees are set by analogues of operational amplifiers. The Paracontrol comprises both analogue and digital systems, and it can be externally adjusted by applying positive and negative voltages. The Paracontrol was tested in real-life experiments with an autonomous mobile robot Emmy, whose favorable/contrary evidences coincide with the values of ultrasonic sensors and distances are represented by continuous values of voltage (Fig. 10.1).

The controller Paracontrol was applied in a series of autonomous mobile robots. In some previous works Da Silva Filho [99] is presented the Emmy autonomous mobile robot. The autonomous mobile robot Emmy consists of a circular mobile platform of aluminum 30 cm in diameter and 60 cm height.

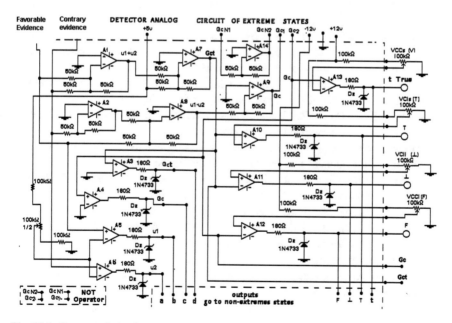

Fig. 10.1 Paracontrol circuit

While moving in a non-structured environment the robot Emmy gets information about presence/absence of obstacles using the sonar system called Parasonic Abe and Da Silva Filho [03]. The Parasonic can detect obstacles in an autonomous mobile robot's path by transforming the distances to the obstacle into electric signals of the continuous voltage ranging from 0 to 5 volts. The Parasonic is basically composed of two ultrasonic sensors of type POLAROID 6500 controlled by an 8051 microcontroller. The 8051 is programmed to carry out synchronization between the measurements of the two sensors and the transformation of the distance into electric voltage. Emmy has suffered improvements, and the 2nd prototype is described in what follows.

The robot Emmy uses the paracontrol system to traffic in non-structured environments avoiding collisions with human beings, objects, walls, tables, etc. The form of reception of information on the obstacles is named non-contact which is the method to obtain and to treat signals from ultra-sonic sensors or optical to avoid collisions.

The system of the robot's control is composed of the Para-sonic, Para-control and supporting circuits. See Fig. 10.2.

- Ultra-Sonic Sensors—The two sensors, of ultra-sonic sound waves accomplish the detection of the distance between the robot and the object through the emission of a pulse train in ultra-sonic sound waves frequency and the return reception of the signal (echo).
- Signals Treatment—The treatment of the captured signals is made through the Para-sonic. The microprocessor is programmed to transform the time elapsed between the emission of the signal and the reception of the echo in an electric signal of the 0–5 volts for the degree of belief, and from 5 to 0 volts for disbelief degree. The width of each voltage is proportional at the time elapsed between the emission of a pulse train and its receivement by the sensor ones.
- Paraconsistent Analysis—The circuit logical controlling paraconsistent makes the logical analysis of the signals according to the logic $E\tau$.
- Codification—The coder circuit changes the binary word of 12 digits in a code of 4 digits to be processed by the personal computer.
- Actions Processing—The microprocessor is programmed conveniently to work the relay in sequences that establish actions for the robot.
- Decodification—The circuit decoder changes the binary word of 4 digits in signals to charge the relay in the programmed paths.
- Power Interface—The power interface circuit potency interface is composed of transistors that amplify the signals making possible the relay's control by digital signals.
- Driving—The relays ON and OFF the motors M_1 and M_2 according to the decoded binary word.
- Motor Drives—The motors M_1 and M_2 move the robot according to the sequence of the relays control.
- Sources—The robot Emmy is supplied by two batteries forming a symmetrical source of the tension of ± 12 Volts DC.

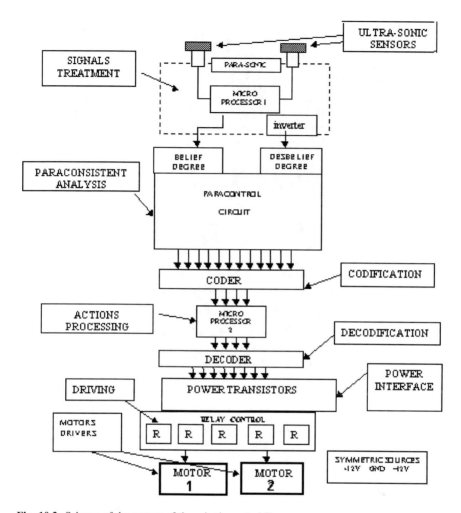

Fig. 10.2 Scheme of the system of the robot's control Emmy

As the project is built in hardware besides the paracontrol it was necessary the installation of components for supporting circuits allowing the resulting signals of the paraconsistent analysis to be addressed and indirectly transformed into action.

In this first prototype of the robot Emmy, it was necessary a coder and a decoder such that the referring signals to the logical states resultants of the paraconsistent analysis had its processing made by a microprocessor of 4 inputs and 4 outputs (Fig. 10.3).

The first robot was dubbed Emmy (in homage to the mathematician Amalie Emmy Noether (1882–1935)) (Abe and Da Silva Filho 2003). Also, it was built a robot based on software using Paralog before mentioned, which was dubbed Sofya

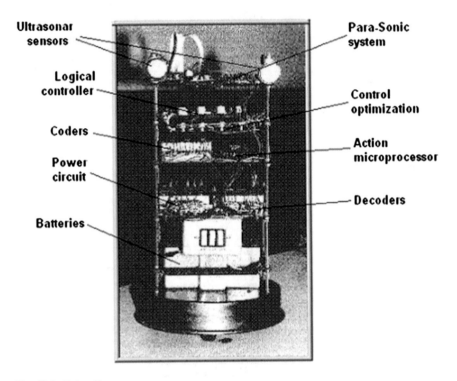

Fig. 10.3 Robot Emmy

(in homage to the mathematician Sofya Vasil'evna Kovalevskaya (=Kowalewskaja) (1850–1891)). Then several other prototypes were made with many improvements. Such robots can deal directly with conflicting and/or paracomplete signals (Torres 2010).

VII. Paraconsistent Artificial Networks

Paraconsistent Artificial Neural Networks—PANN is a new artificial neural network introduced in Da Silva Filho and Abe [01]. Its basis leans on paraconsistent annotated logic $E\tau$ Abe [92], also discussed in another chapter of this book. Roughly speaking, the logic $E\tau$ is a logic that allows contradictions and paracompleteness without trivialization. Also, it is possible to manipulate imprecise concepts in its interior.

10.3   Paraconsistent Knowledge in Distributed Systems

Multi-agents systems are an important topic in AI. The use of modal systems for modelling knowledge and belief has been widely considered in Artificial Intelligence. One interesting approach is due (Fagin et al. 1995).

The essential ideas underlying the systems proposed by (Fagin et al. 1995) can be summarized as follows: $\Box_{iA}$ can be read agent i knows A, i = 1,…, n. Common knowledge and Distributed knowledge are also defined in terms of additional modal operators: $\Box_G$ ("everyone in the group G knows"), $\Box_G^C$ ("it is common knowledge among agents in G"), and $\Box_G^D$ ("it is distributed knowledge among agents in G") for every nonempty subset G of {1,…, n}.

Nevertheless, the most of those proposals use extensions of classical logic or at least part of it, keeping as much as fundamental characteristics of classical logic. When it is taken questions of logical omniscience, one relevant concept that appears is that of contradiction.

The attractiveness of admitting paraconsistency and paracompleteness in the system becomes evident if we observe that some agents can lie or be ignorant about certain propositions: an agent may state both A and ¬A (the negation of A) hold (or that none of A and ¬A hold).

In (Abe and Nakamatsu 2009) it was presented a class of paraconsistent and, in general, paracomplete and non-alethic multimodal systems Jτ which may constitute, for instance, a framework for modelling paraconsistent knowledge.

10.4 A Multi-agent Paraconsistent Framework

In (Prado 1996) it was described a specification and prototype of an annotated paraconsistent logic-based architecture, which integrates various computing systems—planners, databases, vision systems, etc.—of a manufacturing cell.

To make possible the use of such logic in complex application domains (intense information input and critical agent response time), as the manufacture cells, it has been necessary to extend and refine the techniques and concepts of the paraconsistent annotated evidential logic programming and the Amalgam's Knowledge-base.

10.5 Paraconsistent Frame System

In Computer Science, a right solution for a given problem many times depends on a good representation. For most Artificial Intelligence applications, the choice of a knowledge representation is even harder, since the criteria for such choice are less clear.

In (Ávila 1996) it is presented the main features of a paraconsistent inheritance reasoner allowing to deal properly with exceptions and inconsistencies in multiple inheritance frame systems. The paraconsistent inheritance reasoner represents knowledge using paraconsistent frames and infers based on the inconsistency/

under-determinedness degree. This reasoner, being a wide-encompassing one, also allows less complex inheritances to take place.

Furthermore, its main feature is not to eliminate contradictions, *ab initio*.

10.6 Paraconsistent Logic and Non-monotonic Reasoning

There are various intelligent systems including nonmonotonic reasoning in the field of Artificial Intelligence. Each system has different semantics. More than two nonmonotonic reasoning may be required in complex, intelligent systems. It is more desirable to have a common semantics for such nonmonotonic reasoning. We proposed the joint semantics for the nonmonotonic reasoning by annotated logics and annotated logic programs (Nakamatsu et al. 2000).

10.7 Paraconsistent Electronic Circuits

In (Da Silva Filho and Abe 2001) it is presented digital circuits (logical gates COMPLEMENT, AND, OR) inspired in a class of paraconsistent annotated logics Pτ. These circuits allow "inconsistent" signals in a nontrivial manner in their structure.

Such circuits consist of six states; due to the existence of literal operators to each of them, the underlying logic is functionally complete; it is a many-valued and paraconsistent (at least "semantically") logic.

The simulations were made at 50 MHz, 1.2 μm, by using the software AIM-SPICE, version 1.5a. Also, it was presented a paraconsistent analyzer module combining several paraconsistent circuits, as well as a circuit that allows to detect inconsistent signals and gives a non-trivial treatment.

As far as we know, these results seem to be pioneering in using the concept of paraconsistency in the theory of electronic circuits. The applications appear to be large in the horizon: it expands the scope of applications where conflicting signals are common, such as in sensor circuits in robotics, industry automation circuits, race signal control in electronic circuits, and many other fields (Abe and Da Silva Filho 1998).

10.8 Paraconsistent Artificial Neural Networks

Generally speaking, Artificial Neural Network (ANN) can be described as a computational system consisting of a set of highly interconnected processing elements, called artificial neurons, which process information as a response to external stimuli. An artificial neuron is a simplistic representation that emulates the signal integration and threshold firing behaviour of biological neurons using mathematical structures. ANNs are well suited to tackle problems that human beings are good at solving, like prediction and pattern recognition. ANNs have been applied to several branches, among them, in the medical domain for clinical diagnosis, image analysis and interpretation signal analysis and interpretation, and drug development.

Paraconsistent Artificial Neural Network (PANN) (Da Silva Filho et al. 2010) is a new artificial neural network based on paraconsistent annotated logic Eτ (Abe 1992).

The Paraconsistent Artificial Neural Networks were applied in many themes, such as in Biomedicine as prediction of Alzheimer Disease (Lopes et al. 2009, 2011), Cephalometric analysis (Mario et al. 2010), speech disfluency (Abe et al. 2006), numerical characters recognition, robotics (Torres 2010), among others.

10.9 Conclusions

As it can be seen by the previous exposition, the applications of paraconsistent systems have been very fruitful in many aspects, opening new horizons for researching.

The appearance of alternative logics to the classical logic impose us some question to ponder: rationality and logicality coincide? There are in fact logics distinct from classical logic? If there are such alternative logics, there are in consequence distinct rationalities? All these issues occupy philosophers, logicians, scientists in general. Any answer to these questions, we face a true revolution in human thought. A true paradigm is similar like the discovering of the non-Euclidean geometry two centuries ago.

Bibliography

1. Abe, J.M.: Fundamentos da Lógica Anotada (Foundations of Annotated Logic). Doctor Thesis (in Portuguese), Faculdade de Filosofia, Letras e Ciências Humanas (FFLCH), University of São Paulo (USP). São Paulo, Brazil, 98 p (1992)
2. Abe, J.M.: Some aspects of paraconsistent systems and applications. Logique et Anal. **157**, 83–96 (1997)
3. Abe, J.M.: Um Panorama da Lógica Atual. Coleção Cadernos de Estudos e Pesquisas—UNIP, Série: Estudos e Pesquisas, no 1-004/00, ISSN 1517-9230, Universidade Paulista, 28 p (2000)
4. Abe, J.M.: Annotated logics Qτ and model theory, in logic, artificial intelligence, and robotics. In: Proceedings 2nd Congress of Logic Applied to Technology—LAPTEC'2001, Abe, J.M., Da Silva Filho, J.I. (eds.) Frontiers in Artificial Intelligence and Its Applications, IOS Press, Amsterdam, Ohmsha, Tokyo, Editores, vol. 71 (IOS Press), 4 274 90476 8 C3000 (Ohmsha), 1–12, 287 p (2001). ISSN 0922-6389. ISBN 1 58603 206 2
5. Abe, J.M., Da Silva Filho, J.I.: Alexandre SCALZITTI. Introdução à Lógica para a Ciência da Computação. 3a ed. Arte & Ciência Editora. São Paulo, Brasil, 247 p (2002)
6. Abe, J.M., Ortega, N., Mario, M.C., Del Santo Jr., M.: Paraconsistent Artificial Neural Network: an application in cephalometric analysis. In: Lecture Notes in Computer Science, vol. 3.694, Helderberg, pp. 716–723 (2005)
7. Abdel-Kader, M.G., Magdy, G., Dugdale, D., Taylor, P.: Investment Decisions in Advanced Manufacturing Technology—A Fuzzy Set Theory Approach. Ashgate Publishing Ltd., Suffolk, GB (1998)
8. Akama, S., Abe, J.M.: Annotated rules with uncertainty in expert systems. In: IASTED International Applied Informatics, Feb 14–17, 2000, Innsbruck, Austria. Proceedings of IASTED International Applied Informatics. Sponsor: The International Association of Science and Technology for (IASTED), pp. 817–820 (2000a)
9. Akama, S., Abe, J.M.: Fuzzy annotated logics. In: International Conference on Information Processing and Management of Uncertainty in Knowledge-Based Systems., IP MU'2000, 8th. Madrid, Spain, Jul, 3–7, 2000. Organized by: Universidad Politécnica de Madrid. Anais. vol. 1, pp. 504–508 (2000b)
10. Prado, J.P.A.: de. Uma Arquitetura para Inteligência Artificial Distribuída Baseada em Logica Paraconsistente Anotada. Tese apresentada para a obtenção do título de Doutor em Engenharia na Escola Politécnica da Universidade de Sao Paulo (EPUSP). Sao Paulo, Brasil, 217 p (1996)
11. Arruda, A.I.: A survey of paraconsistent logic. In: Arruda, A.I., Chuaqui, R., da Costa, N.C. A. (eds.) Mathematical Logic in Latin. North-Holland Publishing Company, America. Amsterdam, Netherlands, pp. 1–41 (1980)

© Springer International Publishing AG, part of Springer Nature 2018
F. R. de Carvalho and J. M. Abe, *A Paraconsistent Decision-Making Method*,
Smart Innovation, Systems and Technologies 87,
https://doi.org/10.1007/978-3-319-74110-9

12. Avila, B.C.: Uma Abordagem Paraconsistente Baseada em Logica Evidencial para Tratar Exceções em Sistemas de Frames com Múltipla Herança. Tese apresentada para a obtenção do título de Doutor em Engenharia, no Departamento de Engenharia de Computação e Sistemas Digitais da Escola Politécnica da Universidade de Sao Paulo (EPUSP). Sao Paulo, Brasil, 120 p (1996)

13. Berger, J.O.: Statistical Decision Theory and Bayesian Analysis. Springer Series in Statistics, 2nd edn, p. 617. Springer-Verlag, New York, USA (1985)

14. Barreto, M.M.G.: Metodologia Fuzzy para a Construção de Sistemas Especialistas com Bases de Conhecimento Inconsistentes. Tese apresentada para a obtenção do título de Doutor em Ciências em Engenharia Civil na Universidade Federal do Rio de Janeiro (UFRJ). Rio de Janeiro, Brasil, 137 p (1999)

15. Bastos, R., Oliveira, F.M., Oliveira, J.P.M.: Modelagem do processo de tomada de decisão para alocação de recursos. Revista de Administração 33(3), 73–82 (1998)

16. de Bethlem, A.S.: Modelos do processo decisório. Revista de Administração, vol. 22, no. 3. Sao Paulo, Brasil, pp. 27–39 (1987)

17. Bonabeau, E.: Não confie na sua intuição. Harvard Business Review, May, pp. 90–96 (2003)

18. Brunstein, I.: Economia de Empresas: Gestão Econômica de Negócios. la ed. Editora Atlas S.A. Sao Paulo, Brasil, 182 p (2005)

19. Buchanan, I., O'Connell, A.: Uma Breve História da Tomada de Decisões, Harvard Business Review. Brasil, vol. 84, no. 1, Aug (2006)

20. Clemen, R.: Making Hard Decisions: An Introduction to Decision Analysis, 2nd edn. Duxbury Press, Belmont CA (1996)

21. Cassarro, A.C.: Sistemas de Informações para Tomada de Decisões. 3a ed. Editora Pioneira Thomson Learning Ltda. Sao Paulo, Brasil, 129 p (2003)

22. Chalos, Peter: Managing Cost in Today's Manufacturing Environment, p. 283. Prentice Hall Inc, Englewood Cliffs, USA (1992)

23. Clark, J., Harman, M.: On crisis management and rehearsing a plan. Risk Manag. 51(5), 40–43. ABI/INFORM Global, May (2004)

24. Componation, P.J., Sadowski, W.F., Youngblood, A.D.: Aligning strategy and capital allocation decisions: a case study. Eng. Manag. J. 18(1), 24–31. ABI/INFORM Global, Mar (2006)

25. Corner, J., Buchanan, J., Henig, M.: Dynamic decision problem structuring. J. Multi-Criteria Decis. Anal. 10, 129–141 (2001)

26. Correa, H.L.: Administração Geral e Estruturas Organizacionais. Fundação Instituto de Pesquisa Contábeis, Atuariais e Financeiras (FIPECA—FI). Sao Paulo, Brasil, 72 p (2005)

27. de Costa Neto, P.L.O.: Estatistica. 2a ed. Editora Blucher. Sao Paulo, Brasil, 266 pp (2002)

28. de Costa Neto, P.L.O., Bekman, O.R.: Análise Estatistica da Decisão. 2a ed. Editora Blucher. Sao Paulo, Brasil, 148 p (2009)

29. Da Costa, N.C.A., Abe, J.M.: Aspectos Sobre Aplicações dos Sistemas Paraconsistentes. In: Abe J.M. (ed.) Atas do I Congresso de Logica Aplicada a Tecnologia (LAPTEC'2000). Editora Plêiade. Sao Paulo, Brasil (2000). ISBN 85-85795-29-8, 559-571

30. Da Costa, N.C.A., Vago, C., Subrahmanian, V.S.: The Paraconsistent Logics Pτ, in Zeitschrift fur Mathematische Logik und Grundlagen der Mathematik 37, pp. 139–148 (1991)

31. Da Costa, N.C.A., Abe, J.M., Subrahmanian, V.S.: Remarks on annotated logic, Zeitschrift fur Mathematische Logik und Grundlagen der Mathematik 37, pp. 561–570 (1991)

32. Da Costa, N.C.A.: Ensaio Sobre os Fundamentos da Logica. São Paulo, Brasil. Hucitec-Edusp (1980)

33. Da Costa, N.C.A.: Sistemas Formais inconsistentes. Curitiba, Brasil: Editora da Universidade Federal do Parana (UFPR), 66 p (1993)

34. Da Costa, N.C.A.: Conhecimento Científico. 2a ed., Discurso Editorial, Sao Paulo, Brasil, 300 p (1999)

35. Da Costa, N.C.A., Alves, E.H.: A semantical analysis of the Calculi Cn. In: Notre Dame Journal of Formal Logic, v. XVIII, no. 4, October 1977, of University of Notre Dame, Notre Dame, Indiana, USA, pp. 621–630 (1977)

36. Da Costa, N.C.A., Marconi, D.: An overview of paraconsistent logic in the 80's. In: The Journal of Non-Classic Logic, vol. 6, no. 1, pp. 5–32 (1989)

37. Da Costa, N.C.A., Abe, J.M., Murolo, A.C., Da Silva Filho, J.I., Leite, C.F.S.: Logica Paraconsistente Aplicada. Editora Atlas S.A., Sao Paulo, Brasil, 214 p (1999)

38. Da Costa, N.C.A., Abe, J.M.: Algumas Aplicações Recentes dos Sistemas Paraconsistentes em Inteligência Artificial e Robotica. Sao Paulo, Brasil: Institute de Estudos Avan9ados (IEA) da Universidade de Sao Paulo (USP). Sao Paulo, Brasil, 12 p (1999)

39. Da Silva Filho, J.I., Abe, J.M.: Para-fuzzy logic controller—part i: a new method of hybrid control indicated for treatment of inconsistencies designed with the junction of the paraconsistent logic and fuzzy logic. In: Bothe, H., Oja, E., Massad, E., Haefke, C. (eds.) Proceedings of the International ICSC Congress on Computational Intelligence Methods and Applications—CIMA '99, Rochester Institute of Technology, RIT, Rochester, N.Y., USA, ISBN 3-906454-18-5, ICSC Academic Press, International Computer Science Conventions, Canada/Switzerland, pp. 113–120 (1999)

40. Da Silva Filho, J.I., Abe, J.M.: Para-fuzzy logic controller—part ii: a hybrid logical controller indicated for treatment of fuzziness and inconsistencies. In: Bothe, H., Oja, E., Massad, E., Haefke, C. (eds.) Proceedings of the International ICSC Congress on Computational Intelligence Methods and Applications—CIMA '99, Rochester Institute of Technology, RIT, Rochester, N.Y., USA, ICSC Academic Press, International Computer Science Conventions, Canada/Switzerland, pp. 106–112 (1999). ISBN 3-906454-18-5

41. Da Silva Filho, J.I., Abe, J.M.: Paraconsistent analyser module. Int. J. Comput. Anticipatory Syst. 9, 346–352 (2001). ISSN 1373-5411, ISBN 2-9600262-1-7

42. Da Silva Filho, J.I.: Métodos de Aplicações da Lógica Paraconsistente Anotada de Anotação com dois Valores—LPA2v com Construção de Algoritmo e Implementação de Circuitos Eletrônicos. Tese apresentada para a obtenção do título de Doutor em Engenharia, no Departamento de Engenharia de Computação e Sistemas Digitais da Escola Politécnica da Universidade de São Paulo (EPUSP). São Paulo, Brasil, 185 p (1998)

43. Dawes, R.M., Corrigan, B.: Linear models in decision making. Psychol. Bull. 81(2), 93–106 (1974)

44. De Carvalho, F.R.: Logica Paraconsistente Aplicada em Tomadas de Decisão: uma Abordagem para a Administração de Universidades. Editora Aleph. Sao Paulo, Brasil, 120 p (2002)

45. De Carvalho, F.R., Brunstein, I, Abe, J.M.: Paraconsistent annotated logic in the analysis of viability: an approach to product launching. In: Dubois, D.M. (ed.) Sixth International Conference on Computing Anticipatory Systems (CASYS-2003). American Institute of Physics, AIP Conference Proceedings, Springer—Physics & Astronomy, vol. 718, pp. 282–291 (2004). ISBN 0-7354-0198-5, ISSN 0094-243X

46. De Carvalho, F.R., Brunstein, I., Abe, J.M.: Um Estudo de Tomada de Decisão Baseado em Logica Paraconsistente Anotada: Avaliação do Projeto de uma Fábrica. In: Revista Pesquisa & Desenvolvimento Engenharia de Produção, da Universidade Federal de Itajubá, Edição n. 1, dez. pp. 47–62 (2003)

47. De Carvalho, F.R., Brunstein, I., Abe, J.M.: Tomadas de Decisão com Ferramentas da Logica Paraconsistente Anotada. In: Ribeiro, J.L.D., Coppini, N.L., de Souza, L.G.M., Silva, G.P. (eds.) Encontro Nacional de Engenharia de Produção, 23° Ouro Preto, MG, Brasil, 21 a 24 de outubro de 2003. Proceedings. pp. 1–8 (2003b)

48. De Carvalho, F.R., Brunstein, I., Abe, J.M.: Decision making based on paraconsistent annotated logic. In: Nakamatsu, K., Abe, J.M. (eds.) Congress of Logic Applied to Technology (LAPTEC 2005), 5th. Himeji, Japan, April 2–4, 2005. Advances in Logic Based Intelligent Systems: Frontiers in Artificial Intelligence and Applications (Selected papers). IOS Press, Amsterdam, Netherlands, pp. 55–62 (2005a)

49. De Carvalho, F.R., Brunstein, I., Abe, J.M.: Prevision of medical diagnosis based on paraconsistent annotated logic. In: Dubois, D.M. Seventh International Conference on Computing Anticipatory Systems (CASYS—2005). Liege, Belgium, August 8–13, 2005. International Journal of Computing Anticipatory Systems, vol. 18. ISBN 2-930396-04-0, ISSN: 1373-5411, pp. 288–297 (2005b)

50. De Carvalho, F.R.: Aplicação de Logica Paraconsistente Anotada em Tomadas de Decisão na Engenharia de Produção. Tese apresentada a Escola Politécnica da Universidade de Sao Paulo, para a obtenção do título de Doutor em Engenharia. Sao Paulo, Brasil, 349 p (2006)

51. De Carvalho, F.R., Brunstein, I., Abe, J.M.: Decision-making based on paraconsistent annotated logic and statistical method: a comparison. In: Dubois, D.M. (ed.) Eighth International Conference on Computing Anticipatory Systems (CASYS—2007) American Institute of Physics, AIP Conference Proceedings, Springer—Physics & Astronomy, vol. 1.051. ISBN 978-0-7354-0579-0, ISSN: 0094-243X, pp. 195–208 (2008)

52. OP Dias Junior: Decidindo com base em informações imprecisas. Caderno de Pesquisas em Administração 8(4), 69–75 (2001). out/dez

53. Ehrlich, P.J.: Modelos quantitativos de apoio as decisões I. RAE—Revista de Administração de Empresas 36(1), 33–41 (1996a)

54. Ehrlich, P.J.: Modelos quantitativos de apoio as decisões II. RAE—Revista de Administração de Empresas 36(2), 33–41 (1996b)

55. Ecker, J.G., Kupferschmid, M.: Introduction to Operations Research, Krieger Publishing Company (2012). ISBN 0-89464-576-5

56. Fischhoff, B., Phillips, L.D., Lichtenstein, S.: Calibration of probabilities: the state of the art to 1980. In: Kahneman, D., Tversky, A. (eds.) Judgement under Uncertainty: Heuristics and Biases. Cambridge University Press (1982)

57. Gaither, N., Frazier, G.: Administração da Produção e Operações (Production and Operations Management). 8a ed. Editora Pioneira Thomson Learning Ltda. São Paulo, Brasil, 598 p (2001)

58. Gomes, L.F.A.M., Moreira, A.M.M.: Da informação a tomada de decisão: Agregando valor através dos métodos multicritério. COMDEXSUCESU—RIO'98. Rio Centro, Rio de Janeiro, Brasil, em 01/04/1998

59. Gontijo, A.C., Maia, C.S.C.: Tomada de decisão, do modelo racional ao comportamental: uma síntese teórica. Caderno de Pesquisas em Administração, São Paulo 11(4), 13–30 (2004)

60. Goodwin, P., Wright, G.: Decision Analysis for Management Judgment, 3rd edn. Wiley, Chichester (2004). ISBN 0-470-86108-8

61. Gottinger, H.W., Weimann, P.: Intelligent decision support systems. Decis. Support Syst. 8, 317–332 (1992)

62. Gravin, D.A., Roberto, M.A.: What you don't know about making decisions. Harvard Business Review, September pp. 108–116 (2001)

63. Gurgel, F.D.O.A.: Administração do Produto. 2a ed. Editora Atlas S.A. São Paulo, Brasil, 537 p (2001)

64. Hammond, J.S., Keeney, R.L., Raiffa, H.: Smart Choices: A Practical Guide to Making Better Decisions. Harvard Business School Press (1999)

65. Hemsley-Brown, J.: Using research to support management decision making within the field education. Management Decision, vol. 43(5/6), ABI/ INFORM Global, pp. 691–705 (2005)

66. Hilbert, D., Ackermann, W.: Principles of Mathematical Logic, 2nd edn, p. 172. Chelsea Publishing Company, New York, USA (1950)

67. Hillier, F.S., Lieberman, G.L.: Introduction to Operations Research, McGraw-Hill: Boston MA. 8th edn. International edition. ISBN 0-07-321114-1 (2005)

68. Hoffmann, R.: Estatística para Economistas. 4a ed. Editora Thomson. Sao Paulo, Brasil, 432 p (2006)

69. Holloway, C.A.: Decision Making under Uncertainty: Models and Choices. Prentice Hall, Englewood Cliffs (1979)
70. Holtzman, S.: Intelligent Decision Systems. Addison-Wesley (1989)
71. Howard, R.A., Matheson, J.E. (eds.) Readings on the Principles and Applications of Decision Analysis, vol. 2. Strategic Decisions Group, Menlo Park CA (1984)
72. Jarke, M., Radermacher, F.J.: The AI potential of model management and its central role in decision support. Decisi. Support Syst. **4**, 387–404 (1988)
73. Jeusfeld, M.A., Bui, T.X.: Distributed decision support and organizational connectivity: a case study. Decis. Support Syst. **19**, 215–225 (1997)
74. Keeney, R.L.: Value-focused thinking-A Path to Creative Decisionmaking. Harvard University Press. ISBN 0-674-93197-1 (1992)
75. Kleene, S.C.: Introduction to Metamathematics, p. 550. North-Holland Publishing Company, Amsterdam, Netherlands (1952)
76. Lapponi, J.C.: Estatistica usando Excel, 4a edn, p. 476. Editora Elsevier-Campos, Rio de Janeiro, Brasil (2005)
77. Magalhaes, M.N., de Lima, A.C.P.: Noções de Probabilidade e Estatistica. 6a ed. EDUSP, Editora da Universidade de Sao Paulo. Sao Paulo, Brasil, 416 p (2005)
78. Matheson, D., Matheson, J.: The Smart Organization: Creating Value through Strategic R&D. Harvard Business School Press (1998). ISBN 0-87584-765-X
79. de Medeiros, H.R.: Avaliação de modelos matemáticos desenvolvidos para auxiliar a tomada de decisão em sistemas de produção de animais ruminantes no Brasil. 2003. 98 f. Tese apresentada a Escola Superior de Agricultura Luiz de Queiroz (ESALQ), Universidade de São Paulo (USP), para a obtenção do título de Doutor em Engenharia Agrícola. Piracicaba, Brasil (2003)
80. Megginson, L.C., Mosley, D.C., Pietri Jr.P.H.: Administração: Conceitos e Aplicações (Management: Concepts and Applications). Tradução de Maria Isabel Hopp. 4" ed. Editora Harbra Ltda. São Paulo, Brasil, 614 p (1998)
81. Mendelson, Elliott: Introduction to Mathematical Logic, 4th edn, p. 440. Chapman & Hall, New York, NY, USA (1997)
82. Mortari, C.A.: Introdução a Logica. Editora UNESP: Imprensa Oficial do Estado. Sao Paulo, Brasil, 393 p (2001)
83. Nakamatsu, K., Suito, H., Abe, J.M., Suzuki, A.: A theoretical framework for the safety verification of air traffic control by air traffic. In: International Association of Probabilistic Safety Assessment And Management (IAPSAM), Puerto Rico, USA, July, 23–28, 2002. Controllers Based on Extended Vector Annotated Logic Program, PSAM6. Elsevier Science Publishers, Elsevier Science Ltd. (2002)
84. Nakamatsu, K., Abe, J.M.: Railway signal and paraconsistency, advances in logic-based intelligent systems. In: Nakamatsue, K, Abe, J.M. (eds.) Congress of Logic Applied To Technology (LAPTEC 2005), 5th. Himeji, Japan, April, 2–4, 2005. Advances in Logic-Based Intelligent Systems: Frontiers in Artificial Intelligence and Applications (Selected papers). Amsterdam, Netherlands: IOS Press, pp. 220–224 (2005)
85. Negoita, C.V., Ralescu, D.A.: Applications of Fuzzy Sets to Systems Analysis. Wiley, New York, USA (1975)
86. Oliveira, M., Freitas, H.: Seleção de indicadores para a tomada de decisão: A pcrcepcão dos principais intervenientes na construção civil. Revista Eletrônica de Administração (READ), vol. 7(1). Porto Alegre, Brasil, pp. 1–19 (2001)
87. Pal, K., Palmer, A.: A decision-support system for business acquisitions. Decis. Support Syst. **27**, 411–429 (2000).
88. Palma-Dos-reis, A., Zahedi, F.M.: Designing personalized intelligent financial decision support systems. Decis.Support Syst. **26**, 31–47 (1999)
89. Porter, M.: Vantagem competitiva: Criando e sustentando um desempenho superior, tradução de Elizabeth Maria de Pinho Braga; revisão técnica de Jorge A. Garcia Gomez. Editora Campus, Rio de Janeiro, Brasil (1989)

90. Raiffa, H.: Decision Analysis: Introductory Readings on Choices Under Uncertainty. McGraw Hill. ISBN 0-07-052579-X (1997)
91. Roque, V.F., Castro, J.E.E.: Avaliação de risco como ferramenta para auxiliar o sistema de apoio a decisão em indústrias de alimentos. Encontro Nacional de Engenharia de Produção, ENEGEP, 19° Profundão da UFRJ, Rio de Janeiro, Brasil (1999)
92. Saaty, T.L.: Multicriteria Decision Making: The Analytic Hierarchy Process, 2nd edn. RWS Publications, New York, USA (1990)
93. Santos, E. M., Pamplona, E.O.: Captando o valor da flexibilidade gerencial através da teoria das opções reais. Encontro Nacional de Engenharia de Produção, ENEGEP, 21°, Salvador. Anais. Disponível em: http://www.iem.efei.br/edson/download/Art1elieeberenegep01.pdf (2001)
94. Shimizu, T.: Decisões nas Organizações, 2a edn, p. 419. Editora Atlas, Sao Paulo, Brasil (2006)
95. Simon, H.A.: The New Science of Management Decision. Prentice Hall, New York, USA (1960)
96. Skinner, D.: Introduction to Decision Analysis. 2nd edn. Probabilistic (1999). ISBN 0-9647938-3-0
97. Slack, N., Chambers, S., Johnston, R.: Administração da Produção. T ed. Editora Atlas. Sao Paulo, Brasil, 747 p (2002)
98. Smith, J.Q.: Decision Analysis: A Bayesian Approach. Chapman and Hall (1988). ISBN 0-412-27520-1
99. Spiegel, M.R.: Estatistica. 3a edn. Tradução: Pedro Consentino. Makron Books Editora Ltda. Sao Paulo, Brasil, 643 pp (1993)
100. Sycara, K.P.: Machine learning for intelligent support of conflict resolution. Decis. Support Syst. **10**, 121–136 (1993)
101. Taha, H.A.: Operations Research: An Introduction, Prentice Hall. 8th edn (2006). ISBN-10: 0131889230, ISBN 978-0131889231
102. Turban, E., Aronson, J.E.: Decision Support Systems and Intelligent Systems, 6th edn. Prentice International Hall, Hong Kong (2001)
103. Von Neumann, J., Mortgenstern, O.: Theory of Games and Economic Behavior. Wiley, New York, USA (1944)
104. Vetschera, R.: MCView: an integrated graphical system to support multiattribute decision. Decis. Support Syst. **11**, 363–371 (1994)
105. Winston, W.: Operations Research: Applications and Algorithms. 4th. edn. Duxbury Press (2003). ISBN-10: 0534380581, ISBN 978-0534380588
106. Winkler, R.L.: Introduction to Bayesian Inference and Decision, 2nd edn. Probabilistic (2003). ISBN 0-9647938-4-9
107. Witold, P. (with a foreword by Lotfi A. Zadeh). Fuzzy Sets Engineering. CRC Press Inc., Florida, USA, 332 p (1995). ISBN 0-8493-9402-3
108. Witold, P., Gomide, F.: An introduction to fuzzy sets. MIT Press, Massachusetts, USA, ISBN 0-262-16171-0, 465 p (1998)
109. Woiler, S., Mathias, W.F.: Projetos: Planejamento, Elaboração e Analise (Projects: Planning, Elaboration and Analysis). Editora Atlas. Sao Paulo, Brasil, 294 p (1996)
110. Wu, J.-H., Doong, H.-S., Lee, C.-C., Hsia, T.-C., Liang, T.-P.: A methodology for designing form-based decision support systems. Decis. Support Syst. **36**, 313–335 (2004)
111. Xu, L., Li, Z., Li, S., Tang, F.: A decision support system for product design in concurrent engineering. Decis. Support Syst. In Press (2004)
112. Zadeh, L.A.: Outline of a New Approach to the Analysis of Complex Systems and Decision Processes—IEEE Transaction on Systems, Mam and Cybernetics, vol. SMC-3, no. 1, pp. 28–44 (1973)
113. Abe, J.M., S. Akama, K. Nakamatsu,: Introduction to Annotated Logics—Foundations for Paracomplete and Paraconsistent Reasoning, Series: Intelligent Systems Reference Library, Vol. 88, Springer International Publishing (2015)

114. Abe, J.M., K. Nakamatsu & S. Akama: Two Applications of Paraconsistent Logical Controller, in New Directions in Intelligent Interactive Multimedia, serie Studies in Computational Intelligence, Springer Berlin / Heidelberg, Vol. 142, 249–254 (2008)

115. Abe, J.M., da Silva Filho, J.I.: Manipulating Conflicts and Uncertainties in Robotics. Multiple-Valued Logic and Soft Computing **9**, 147–169 (2003)

116. Abe, J.M., Nakamatsu, K.: Multi-agent Systems and Paraconsistent Knowledge. In: Nguyen, N.T., Jain, L.C. (eds.) Knowledge Processing and Decision Making in Agent-Based Systems, Book series Studies in Computational Intelligence, vol. 167, VIII, 400 p. 92 illus., Springer-Verlag, 101–121 (2009)

117. Abe, J.M., Da Silva Filho, J.I.: Inconsistency and Electronic Circuits. In: Alpaydin, E. (ed.) Proceedings of the International ICSC Symposium on Engineering of Intelligent Systems (EIS'98), vol. 3, Artificial Intelligence, ICSC Academic Press International Computer Science Conventions Canada/Switzerland, 191–197 (1998)

118. Abe, J.M., Prado, J.C.A., Nakamatsu, K.: Paraconsistent Artificial Neural Network: Applicability in Computer Analysis of Speech Productions, Lecture Notes in Computer Science 4252, pp. 844–850, Springer (2006)

119. Abe, J.M.: Paraconsistent Intelligent Based-Systems: New Trends in the Applications of Paraconsistency, editor, Book Series: "Intelligent Systems Reference Library", Springer-Verlag, Vol. 94, Germany (2015)

120. Akama, S.: Towards Paraconsistent Engineering, Intelligent Systems Reference Library, Vol. 110, Springer International Publishing (2016)

121. Da Costa, N.C.A., Prado, J.P.A., Abe, J.M., Ávila, B.C., Rillo, M.: Paralog: Um Prolog Paraconsistente baseado em Lógica Anotada, Coleção Documentos, Série Lógica e Teoria da Ciência, IEA-USP, no 18, 21 p (1995)

122. Da Silva Filho, J.I., Abe, J.M.: Paraconsistent electronic circuits. Int. J. Comput. Anticipatory Syst. **9**, 337–345 (2001)

123. Da Silva Filho, J.I., Torres, G.L., Abe, J.M.: Uncertainty Treatment Using Paraconsistent Logic—Introducing Paraconsistent Artificial Neural Networks, IOS Press, Holanda, vol. 211, 328 pp (2010)

124. Fagin, R., Halpern, J.Y., Moses, Y., Vardi, M.Y.: Reasoning about Knowledge. The MIT Press, London (1995)

125. Lopes, H.F.S., Abe, J.M., Anghinah, R.: Application of Paraconsistent Artificial Neural Networks as a Method of Aid in the Diagnosis of Alzheimer Disease, Journal of Medical Systems, Springer-Netherlands, pp. 1–9 (2009)

126. Lopes, H.F.S., Abe, J.M., Kanda, P.A.M., Machado, S., Velasques, B., Ribeiro, P., Basile, L.F.H., Nitrini, R., Anghinah, R.: Improved application of paraconsistent artificial neural networks in diagnosis of alzheimer's disease. Am. J. Neurosci. **2**(1), 54–64 (2011). Science Publications

127. Mario M.C., Abe, J.M., Ortega, N., Del Santo Jr., M.: Paraconsistent Artificial Neural Network as Auxiliary in Cephalometric Diagnosis. Artif. Organs **34**(7), 215–221 (2010) Wiley Interscience

128. Nakamatsu, K., Abe, J.M., Suzuki, A.: Annotated semantics for defeasible deontic reasoning, rough sets and current trends in computing, lecture notes in artificial intelligence series. Springer-Verlag **2005**, 470–478 (2000)

129. Prado, J.P.A.: Uma Arquitetura em IA Baseada em Lógica Paraconsistente, Ph.D. Thesis (in Portuguese), University of São Paulo (1996)

130. Subrahmanian, V.S.: On the semantics of quantitative logic programs. In: Proceedings 4th IEEE Symposium on Logic Programming, Computer Society Press, Washington D.C., 173–182 (1987)

131. Torres, C.R.: Sistema Inteligente Baseado na Lógica Paraconsistente Anotada Evidencial Eτ para Controle e Navegação de Robôs Móveis Autônomos em um Ambiente Não-estruturado, PhD. Thesis (in Portuguese), Federal University of Itajuba, Brazil (2010)

Printed in the United States
By Bookmasters